Electrospinning and Electrospraying - Techniques and Applications

Edited by Sajjad Haider and Adnan Haider

Published in London, United Kingdom

IntechOpen

Supporting open minds since 2005

Electrospinning and Electrospraying - Techniques and Applications
http://dx.doi.org/10.5772/intechopen.77414
Edited by Sajjad Haider and Adnan Haider

Contributors
Olga Starinova, Andrey Shornikov, Elizaveta Nikolaeva, Marilena Vlachou, Angeliki Siamidi, Sotiria Kyriakou, Kim Ick Soo, Amal Ali Elkordy, Ijeoma Abraham, Eman Ali Elkordy, Rita Haj Ahmad, Zeeshan Ahmad, Sajjad Haider

First published in London, United Kingdom, 2019 by IntechOpen
IntechOpen is the global imprint of INTECHOPEN LIMITED, registered in England and Wales, registration number: 11086078, 7th floor, 10 Lower Thames Street, London, EC3R 6AF, United Kingdom
Printed in Croatia

British Library Cataloguing-in-Publication Data
A catalogue record for this book is available from the British Library

Additional hard and PDF copies can be obtained from orders@intechopen.com

Electrospinning and Electrospraying - Techniques and Applications
Edited by Sajjad Haider and Adnan Haider
p. cm.
Print ISBN 978-1-78984-700-0
Online ISBN 978-1-78984-701-7
eBook (PDF) ISBN 978-1-83968-113-4

For EU product safety concerns:
IN TECH d.o.o., Prolaz Marije Krucifikse Kozulić 3, 51000 Rijeka, Croatia, info@intechopen.com or visit our website at intechopen.com.

Meet the editors

Dr. Sajjad Haider has been an associate professor at the Chemical Engineering Department, King Saud University, Riyadh, Saudi Arabia, since May 2009. He received his MSc degree in 1999 and his MPhil degree in 2004 from the Institute of Chemical Sciences, University of Peshawar, KPK, Pakistan, and PhD degree in 2009 from the Department of Polymer Science and Engineering, Kyungpook National University, Daegu, South Korea. His research work focuses on the development of carbon nanotubes and biopolymer composites, polymer hydrogel, and electrospun nanofibers for energy storage, and environmental and biomedical applications.

Dr. Adnan Haider is an assistant professor at the Department of Chemistry, Kohat University of Science and Technology. He received his MSc degree in 2010 from the Kohat University of Science and Technology, KPK, Pakistan; his PhD degree in 2016 from the Department of Polymer Science and Engineering, Kyungpook National University, Daegu, South Korea, and post-doctoral research in 2017 from the Department of Nano, Medical and Polymer Materials, School of Chemical Engineering, Yeungnam University, South Korea. His research work focuses on the development of scaffolds using polymer hydrogel and electrospun nanofibers for biomedical (tissue regeneration and drug delivery) and environmental applications.

Contents

Preface

As progress in nanotechnology was being made, researchers developed new techniques for the fabrication of nanomaterials with a variety of morphologies. These morphologies were usually aimed at a specific application. Among the various techniques reported in the literature, electrospinning and electrospraying have since gathered significance. Electrospinning and electrospraying are "sister" technologies used to produce polymer-based nanomaterials. Although these technologies use almost the same apparatus and share the same principle (electro-hydrodynamic), their final products are different. The product of electrospinning is the nanofiber whereas the product of electrospraying is the nanobead. Both of these electrodynamic techniques have been receiving increasing attention not only in the scientific community but also in industry. The products of these technologies could be used both separately and together in a wide range of applications. For example, the nanocomposite (combined state) nonwoven fabrics could be used as a filtration mat with improved efficiency due to incorporated catalytic material and electrodes of high specific surface area for fuel cells. Both these techniques have resulted in positive impacts in the biomedicine field and are being tested for food processing and preservation, which is a less explored area.

This book intends to provide the reader with a comprehensive overview of the current advances in electrospinning and electrsospraying technologies and their applications. The book will be beneficial not only to beginners but will also attract the attention of experts in the field.

Sajjad Haider
King Saud University,
Saudi Arabia

Adnan Haider
Kohat University of Science and Technology,
Pakistan

Section 1

Introduction

Chapter 1

Electrohydrodynamic Processes and Their Affecting Parameters

Sajjad Haider, Adnan Haider, Abdulaziz A. Alghyamah, Rawaiz Khan, Waheed A. Almasry and Naeem Khan

Abstract

Electrohydrodynamic processes such as electrospinning and electrospraying are simple, flexible, and cost-effective. Both processes use electrically charged jet of polymer solution for the fabrication of micro- or nanofibers and micro- or nanoparticle. Both of these electrodynamic techniques have been receiving increasing attention not only in the scientific community but also in industry. These fibers and particles offer several morphological and functional features that are suitable for tissue engineering in biomedical applications. The main apparatus used for both of these processes is almost the same. Both need electric voltage to induce charge on the droplet, which at optimized electric field leads to micro- or nanofibers and micro- or nanoparticles. Rayleigh in 1882, for the first time, theoretically estimated the maximum amount of charge that a liquid droplet could carry to change in a jet. This theory is now known as the "Rayleigh limit." He predicted that a droplet on reaching Rayleigh limit would move as fine jets of liquid. More than 100 years later, Rayleigh limit theory was confirmed experimentally. Beside electric field there are other operating and solution parameters that need to be optimized before we obtain a desire product.

Keywords: electrohydrodynamic processes, cost effective, micro and nanofibers, micro and nanoparticles

1. Electrospinning

Since the late twentieth century, electrospinning, a technology used for the fabrication of nanofibers, has been receiving increasing attention not only in the scientific community but also in industry. With this technique fibrous material was conveniently prepared with fascinating properties such as ultrafine diameters in the range of 10–500 nanometer (nm), high permeability, porosity, surface area per unit mass, and small inter-fibrous pore size [1–3]. It is, therefore, considered to be a vital scientific and commercial venture with global economic benefits. This technology traced its legacy back to 1902 [4, 5] and 1934 [6], when the very first patents entitled "Apparatus for Electrical Dispersion of Fluids" and "Process and Apparatus for Preparing Artificial Over and Done Threads from Electrically Dispersed Fluids" were registered. Since then, this process was largely ignored until the 1990s. However, with the recent revelations by researchers who are associated with the field of nanotechnology mainly nanofibers, investigations into the preparation of nanofibers using the aforementioned technique picked up momentum [7].

In the literature, various techniques are reported for the fabrication of nanomaterials. These include drawing-processing, template-assisted synthesis, self-assembly, solvent casting, phase separation, and electrospinning [8–10]. With advancement of research in nanotechnology, particularly associated with nanoparticles, nanostructures, and more explicitly with the preparation of scaffolds, electrospinning emerged as a highly developed and frequently used technique/process. This process is favored over other techniques such as solvent casting and phase separation, since nanofibers have high surface area to volume ratio and inter-/intra-fibrous pores. In addition to the previously mentioned properties, it has the advantages of being easy to use and have a low processing cost. The growing literature on electrospinning has helped this technique to pave the way for advancements in areas like environmental protection, bioengineering, electronics, and catalysis [11–13]. The capability of this technique to produce nanomaterials from numerous raw materials ranging from simple natural polymers to complex materials such composites has attracted a large number of researchers. For example, scientists have reported a wide range of applications of electrospun nanofibers in the protection of the environment, most importantly in water and air filtration. Subramanian et al. [14] and Feng et al. [15] emphasized the importance of using electrospun nanofibers for the removal of contaminants from water, focusing on the application of electrospun nanofibers in nanofiltration.

Furthermore, polymers with a piezoelectric effect (i.e., polyvinylidene fluoride) can be subjected to an electrospinning technique for the fabrication of nanofiber scaffolds with a piezoelectric effect. Apart from the use of nanofibers in scaffolds with piezoelectric effect, they can also be used for producing high-surface-area nanosensors. Huang et al. suggested that scaffolds of the copolymer poly(lactide-co-glycolide) (PLGA) could be used as sensing tools in both chemical and biological fields [16]. Investigators have also highlighted that sensors prepared of nanofibers might demonstrate improved sensing capabilities for chemicals such as 2,4-dinitrotulene DNT, mercury (Hg), and ferric (Fe^{+3}) ions as compared to orthodox thin film.

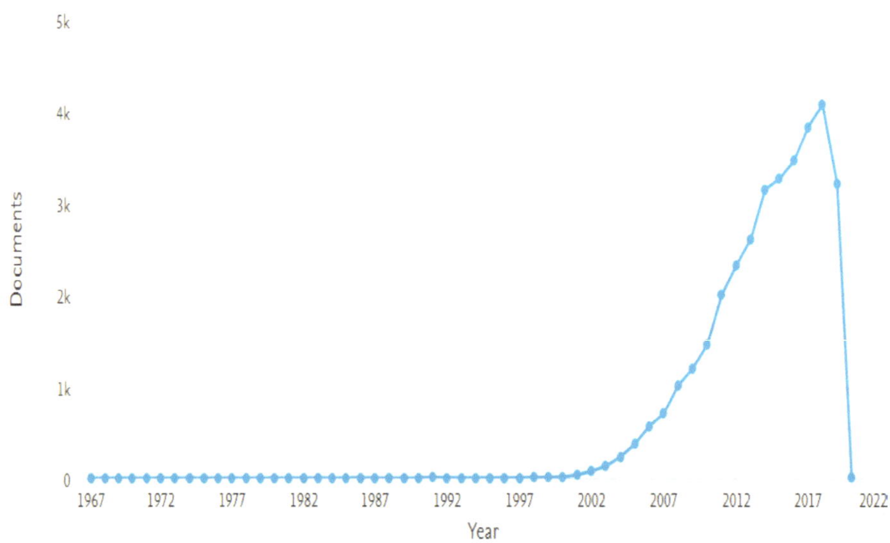

Figure 1.
Scopus data showing year-wise publications in the area of electrospinning. The data is taken on February 09, 2019.

Furthermore, fluorescent polymer electrospun nanofiber optical sensors also showed high sensitivity [17]. Ultrafine electrospun nanofiber scaffolds can also be applied in the preparation of nanotubes, which are of prime importance in various industries [18]. To prepare nanotubes, electrospun nanofibers are coated with the raw material of the nanotubes, and upon the evaporation of a solvent via solvent evaporation or the thermal degradation of the polymers, the nanotubes are synthesized. Bognitzki et al. adopted a physical and chemical vapor deposition technique using poly (L-lactide)(PLLA) nanofibers as a template and synthesized polymer composite nanotubes comprised of poly(p-xylene)/aluminum metal with a thickness of 0.1–1 mm [19]. Hou et al. prepared nanotubes with an even smaller diameter by adopting the same technique using poly(L-lactide)(PLA) and poly(tetramethylene adipamide)(PA) as templates [16, 20].

Electrospun nanofibers have also played a pivotal role in the area of biomaterials. The importance of electrospun nanofibers in the biomedical field can be determined from the fact that numerous articles are being published every

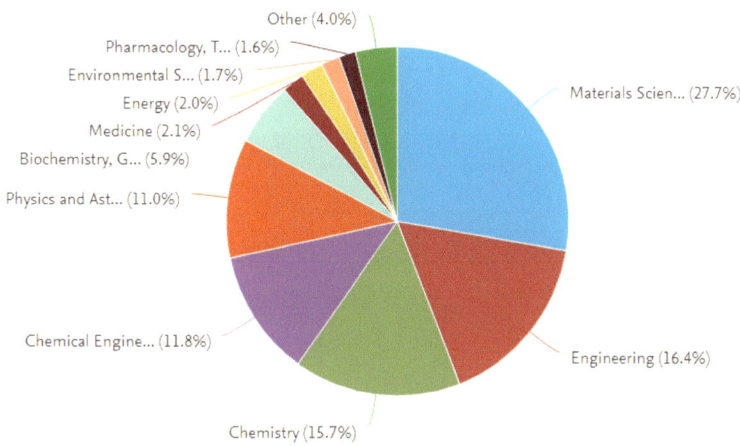

Figure 2.
Showing Scopus data subject wise where electrospinning has been used. The data is taken on February 09, 2019.

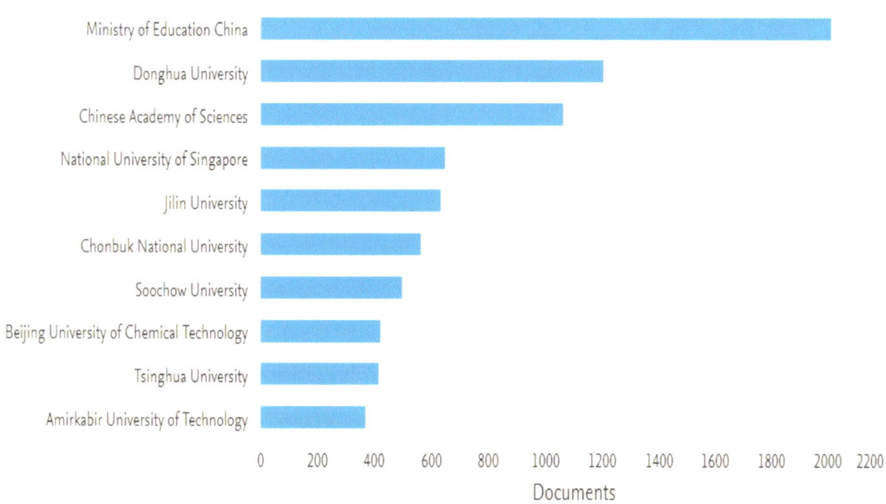

Figure 3.
Showing Scopus data of the institutions, which are working in area of electrospinning. The data is taken on February 09, 2019.

year on a regular basis in high-quality journals, highlighting their significance in biomedical engineering. Nanofiber scaffolds were also functionalized with desired triggering groups. These triggering groups performed a significant role in providing conducive atmosphere to the cells, which resulted in their enhanced

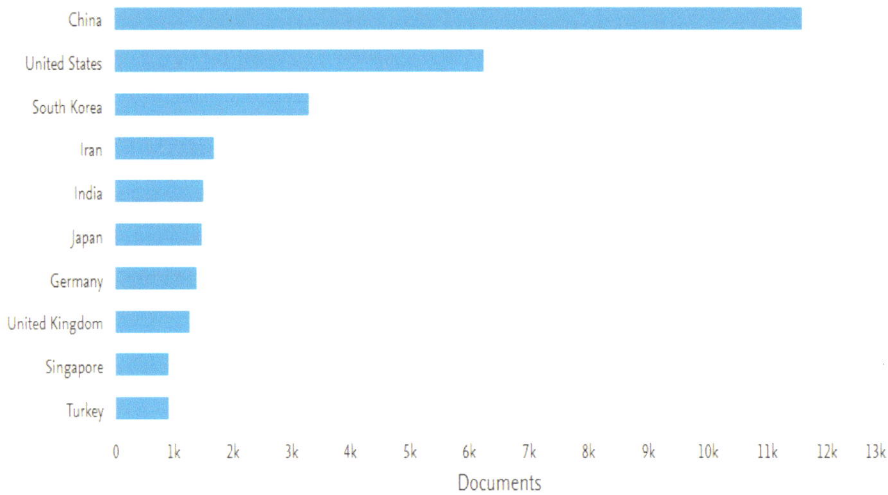

Figure 4.
Country-wise Scopus data of electrospinning. The data is taken on February 09, 2019.

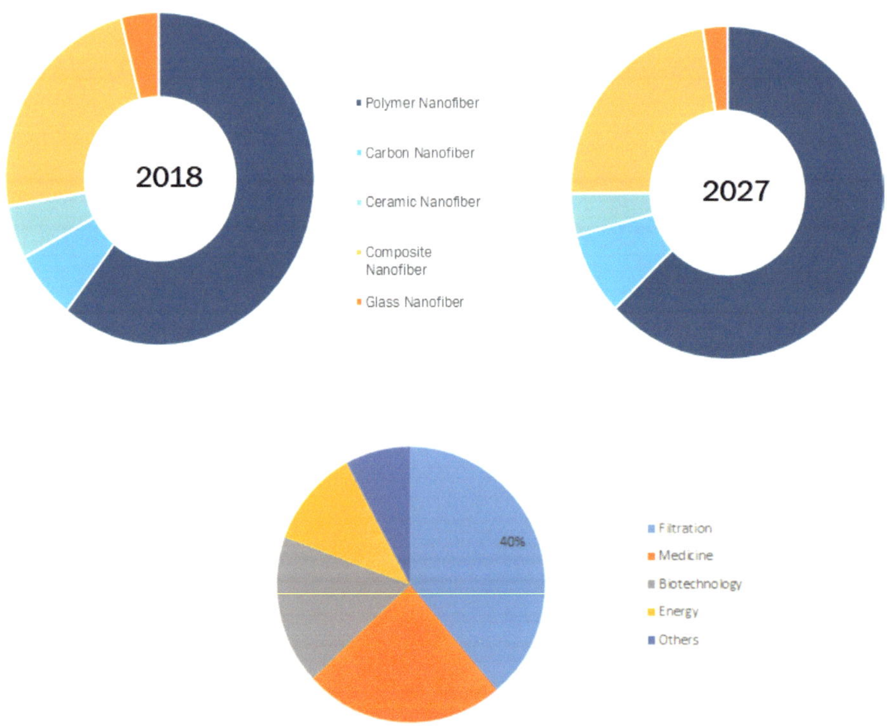

Figure 5.
Nanofiber market to 2025—Global analysis and forecasts by material (polymer nanofiber, carbon nanofiber, ceramic nanofiber, composite nanofiber, and glass nanofiber) and application (energy, medical life science and pharmaceutical, and chemical and environmental and defense and security, consumer, electronics, https://www.theinsightpartners.com/reports/nanofiber-market). Polymer nanofiber market, research report—Forecast to 2023 https://www.marketresearchfuture.com/reports/polymer-nanofiber-market-4416.

anchoring, proliferation, and differentiation. For example, collagen fibril has been known to enhance the interaction between cells and scaffolds. Similarly, electrospun nanofiber scaffolds are also used as a drug delivery carrier for carrying drugs to their target sites [21, 22].

Nanofibers, until now, have been produced using electrospinning from nearly 100 diverse polymers of natural and synthetic origins. All polymers have been electrospun via solvent and melt spinning. Even with the persistent use of this technique, the understanding of the basics still need to be acquired. **Figures 1–5** show the research interest in the area of electrospinning and the market interest in the electrospun product. More than 200 universities and research institutes worldwide (some with high publications are shown in **Figure 3**) are still studying a variety of the electrospinning processes, their various aspects, and the nanofibers produced. The market of electrospun product is expected to increase mostly in filtration and medical fields by 2025 (**Figure 5**).

2. Electrospinning and its mechanism

Much research has been done on the electrospinning technique. Based on the literature, the fundamental setup used for electrospinning (**Figure 6a**) consists of mainly four parts: a glass syringe (holding solution), needle (metallic), applied voltage, and collector (metallic, with a variable morphology). The process commences, when electric charges, produced on the needle due to the applied voltage, transfer into the polymer solution via the metallic needle [22]. These charges cause instability within the polymer droplet. The repulsion of charges generates a force opposite to surface tension. These forces cause the polymer solution to move in the direction of the electric field (**Figure 6b**). An increase in the electric field forces the spherical droplet to distort and adopt a conical shape. At this phase, fine nanofibers (nano to micro in diameters) appear from the conical droplet called Taylor cone, which are whippingly collected on the collector and kept at an optimized distance. A steady charge jet could only form, when the polymer solution possesses adequate cohesive force. This whipping of the fibers permits the polymer chains to stretch

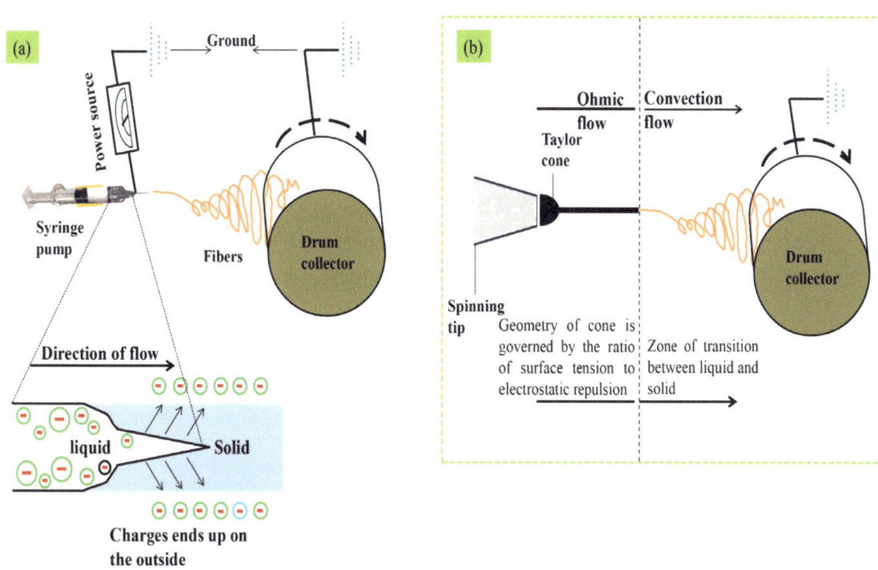

Figure 6.
Schematic depicting (a) electrospinning setup, and (b) phenomenon of electrospinning.

and slide past each other in the solution, which leads to the fabrication of fibers with diameters as mentioned above [23, 24].

3. Effects of parameters on electrospinning

There are various factors that affect the electrospinning process. These factors are classified as electrospinning parameters and solution parameters. The electrospinning parameters include the applied electric field, distance between the needle and collector, flow rate, and needle diameter. The solution parameters include the solvent, polymer concentration, and viscosity. All of these parameters directly affect the generation of smooth and bead-free electrospun fibers. Therefore, to gain a better understanding of the electrospinning technique and fabrication of polymeric nanofibers, it is essential to thoroughly understand the effects of all of these governing parameters. Detailed information highlighting the role of electrospinning and the effects of the solution parameters on the morphology of electrospun polymeric nanofibers are summarized below.

3.1 Effect of changes in applied voltage

Generally, it is a known fact that the flow of current from a high-voltage power supply into a solution via a metallic needle will cause a spherical droplet to deform into a Taylor cone and form ultrafine nanofibers at a critical voltage (**Figure 7a–f**). This critical value of applied voltage varies for different polymers systems. The creation of nanofibers with smaller diameter as the applied voltage is increased could be ascribed to the stretching of the polymer solution and charge repulsion in the polymer jet [25]. An increased applied voltage further than the critical value

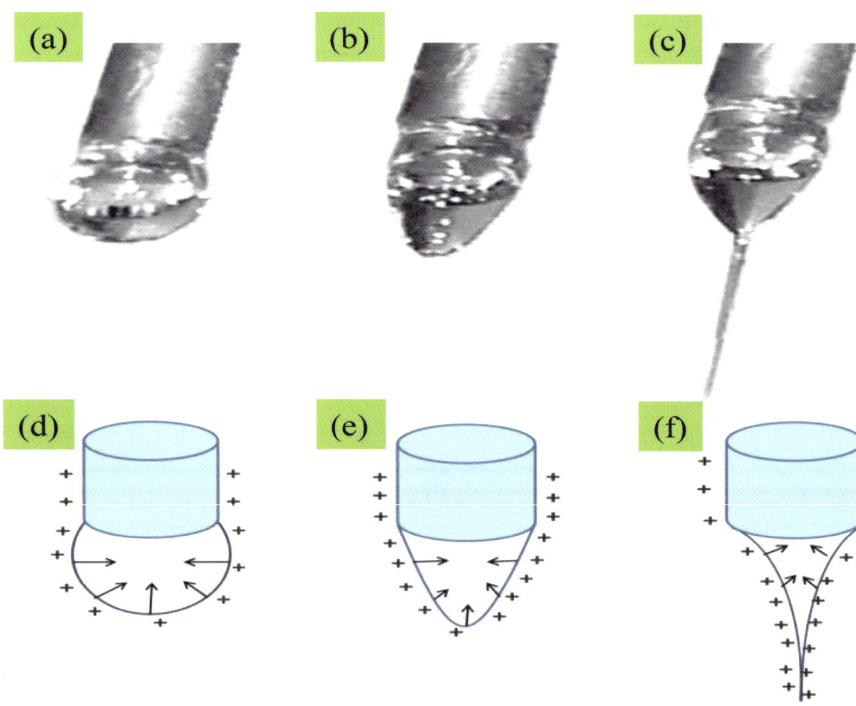

Figure 7.
The images showing the droplet deformation and eventual fiber ejection with an increase in electric field on droplet and effect of charges on polymeric solutions [28].

while keeping the flow rate constant will lead to the creation of beads or beaded nanofibers. The creation of beads or beaded nanofibers with an increased applied voltage is ascribed to the decrease in size of the Taylor cone and increase in the jet velocity. Deitzel et al. reported bead formation with an increase in the applied voltage using poly(ethylene oxide) (PEO)/water. Similar results were also reported by Meechaisue et al. and Zong et al. [26]. Furthermore, the diameter of the nanofibers was also reported to increase with an increase in the applied voltage. This increase in the diameter was attributed to an increase in the jet length with the applied voltage (**Figure** 7) [27].

4. Effect of solution flow rate

The movement of solution through the tip of the needle controls the morphology of the nanofibers. Uniform beadless nanofibers could be fabricated via a critical flow rate. The flow rate varies for different polymers systems. An increase in the flow rate further than the critical value results in the creation of beads. For instance, the increase in flow rate of polystyrene (PS) to 0.10 mL/min led to bead creation. However, when the flow rate was reduced to 0.07 ml/min, bead-free, narrow-diameter electrospun fibers were formed. Because an increase and the decrease in the flow rate affect the nanofiber diameter, a minimum flow rate is preferred to maintain a balance between the leaving polymeric solution and replacement of that solution with a new one during jet formation [29, 30]. This will also permit the formation of a steady cone jet and sometimes a receded jet: a jet that appears from the inside of the needle with no apparent droplet or cone. Receded jets are not steady jets, and during the process, these jets are unceasingly substituted by cone jets. This phenomenon results in the formation of nanofibers with wide range diameters (**Figure 8a–h**) [31]. Numerous research groups have studied the effect of the flow rate on the morphology and diameter of nanofibers. For instance, Zong et al. revealed that smaller diameter nanofibers can be prepared using a low flow rate, whereas a high flow rate will not only yield high-diameter nanofibers, but the bead formed will also have a high diameter [32]. Megelski et al. showed an increase in the pore size and fiber diameter of PS nanofibers by increasing the flow rate of the polymeric solution. They also concluded that fibers with beads were formed at a high flow rate as a result of the incomplete drying of the nanofibers jet during its flight between the needle tip and metallic collector [29]. In addition to bead formation, in some cases, at an elevated flow rate, ribbonlike defects [29] and unspun droplets (**Figure 8e**) have also been reported in the literature [31]. The creation of beads and ribbonlike structures with an increased flow rate was mainly ascribed to the low stretching of the solution in the flight between the needle and metallic collector and the non-evaporation of the solvent. Similar effect could also be ascribed to an increase in diameter of the nanofibers with an increase in the flow rate [29]. The presence of the unspun droplets is attributed to the influence of the gravitational force [31]. Another important factor that may cause defects in the nanofiber structure is the surface charge density. Any change in the surface charge density may also affect the morphology of the nanofibers. For example, Theron et al. found that flow rate and electric current are directly linked to each other. His group investigated the effects of the flow rate and surface charge density using PEO, polyacrylic acid (PAA), polyvinyl alcohol (PVA), polyurethane (PU), and polycaprolactone (PCL). They noted that, in the case of PEO, an increase in the flow rate increased the electric current and decreased the surface charge density at the same time. A decreased surface charge density permits the merger of nanofibers in their flight to the collector. This merger of nanofibers helps in the creation of garlands [33, 34].

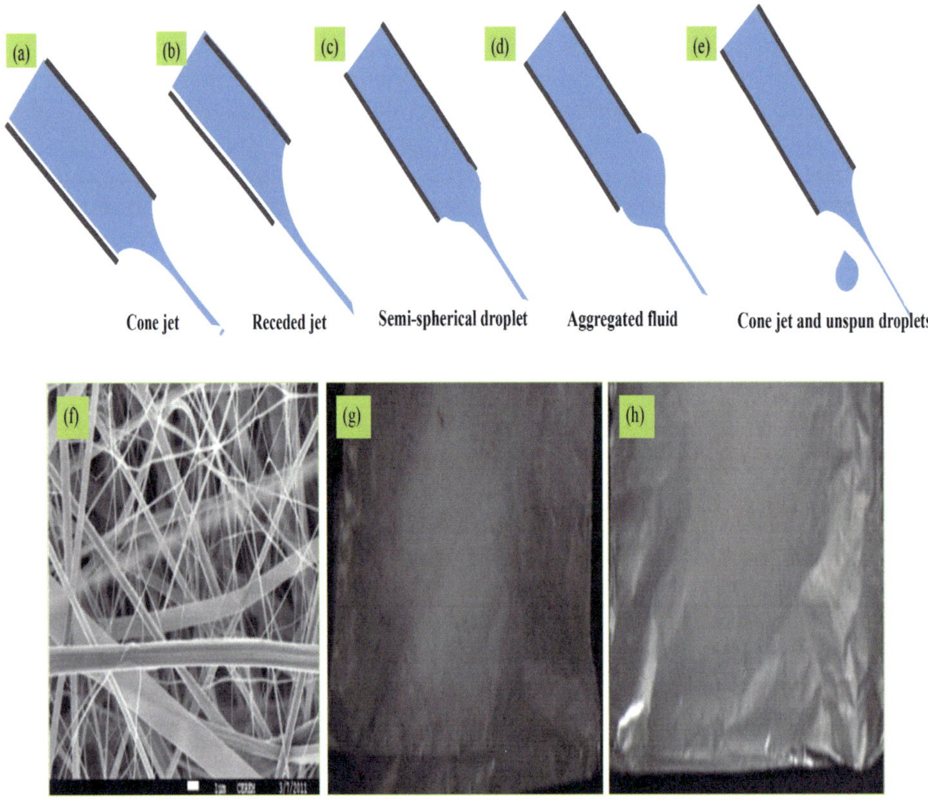

Figure 8.
Formation of various jets with increasing flow rate; here, (f–h) show images of electrospun fibers on aluminum foil [31].

4.1 Effects of polymer concentration and solution viscosity

Electrospinning counts on the uniaxial stretching of a charged jet. The stretching of the charged jet is considerably affected if the solution concentration is changed. For instance, when the concentration is low, the applied electric field and surface tension force the entangled polymer chains to rupture into fragments before reaching the collector [23, 35] . These fragments then lead to the creation of beads or beaded nanofibers. Increased concentration of the solution, however, leads to an increase in the viscosity, which subsequently increases the chain entanglement. These entanglements then overcome the surface tension and eventually result in uniform and beadless nanofibers. Moreover, increasing the concentration further than a critical value impedes the flow of the solution through the needle tip. In simple words viscous solution partially blocks and dries at the needle tip. This eventually leads to defective and beaded nanofibers [23]. The morphologies of the beads depict an interesting shape change from a round droplet-like shape (with low-viscosity solutions) to a stretched droplet or ellipse to smooth fibers (with sufficient viscosity) as the solution viscosity changes (**Figure 9a–h**). This effect of the concentration/viscosity on the morphology of the nanofibers was also reported by Doshi et al. Working with PEO, they concluded that the optimum viscosity for the generation of electrospun nanofibers is 800–4000 cp [7]. In addition to the work of Doshi et al., an experiment on a polyacrylonitrile (PAN) polymer solution showed that smooth electrospun nanofibers could be prepared when the viscosity of the solution was kept at 1.7–215 cp. Hence, it can be concluded that in addition to the

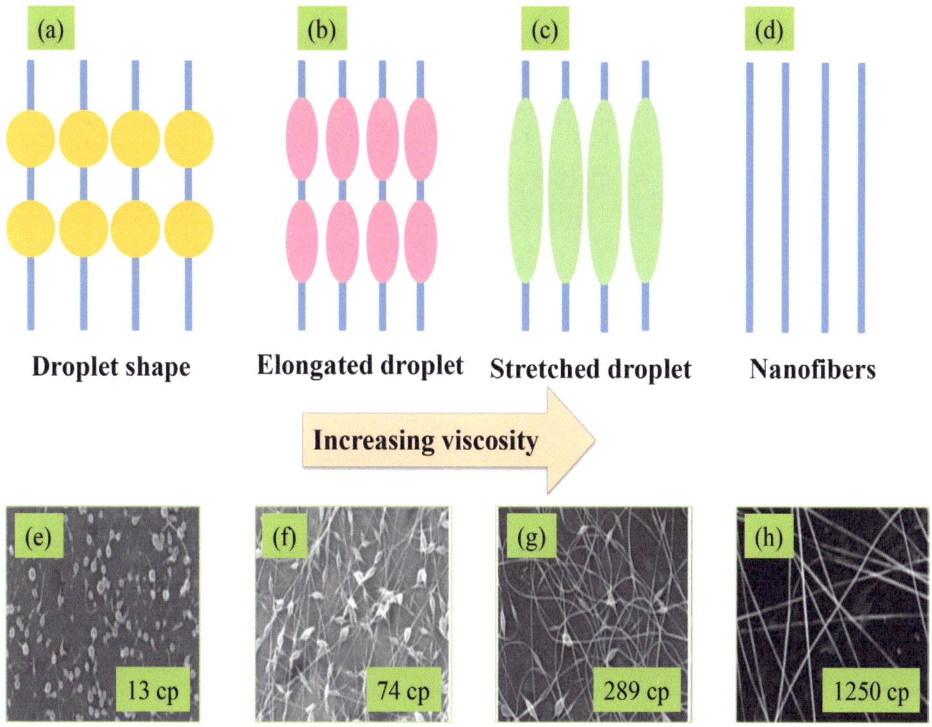

Figure 9.
Variation in morphology of electrospun nanofibers with viscosity: (a–d) schematic and (e–h) SEM micrographs [36, 37].

electrospinning parameters, the determination of the critical value of the concentration/viscosity is also essential to obtain beadless nanofibers [27].

4.2 Role of solvent in electrospinning

The selection of solvent is one of the key factors for the fabrication of smooth and beadless electrospun nanofibers. Usually two things need to be kept in mind before selecting the solvent. First, the preferred solvents for electrospinning process have polymers that are completely soluble. Second, the solvent should have a moderate boiling point. Its boiling point gives an idea about the volatility of a solvent. Mostly volatile solvents are desired as their high evaporation rates boost the fabrication of nanofibers. However, very highly volatile solvents are commonly avoided since their high evaporation rate forces the drying of jet at the needle tip. This drying blocks the needle tip and thus hampers the electrospinning process. Likewise, low-volatile solvents are also avoided since their low evaporation inhibits their drying in the flight. The deposition of solvent-containing nanofibers on the collector will cause the formation of beaded nanofibers [25, 38]. Numerous research groups have studied the effects of the solvent and solvent system on the morphology of nanofibers (**Figure 10a–f**) [39] and concluded that similar to the applied voltage, solvent also affects the polymer system [40]. Moreover, solvents also play a vibrant role in the fabrication of highly porous nanofibers. This may happen when a polymer is dissolved in two solvents, where one act as solvents and the other as a non-solvent. The difference in the evaporation rates of the solvent and non-solvent causes phase separation, which results in the creation of pores in nanofibers (**Figure 10f**) [25]. Similar results were also reported by Zhang et al. [41]. Megelski et al. prepared porous nanofibers by varying the ratio of tetrahydrofuran (THF)

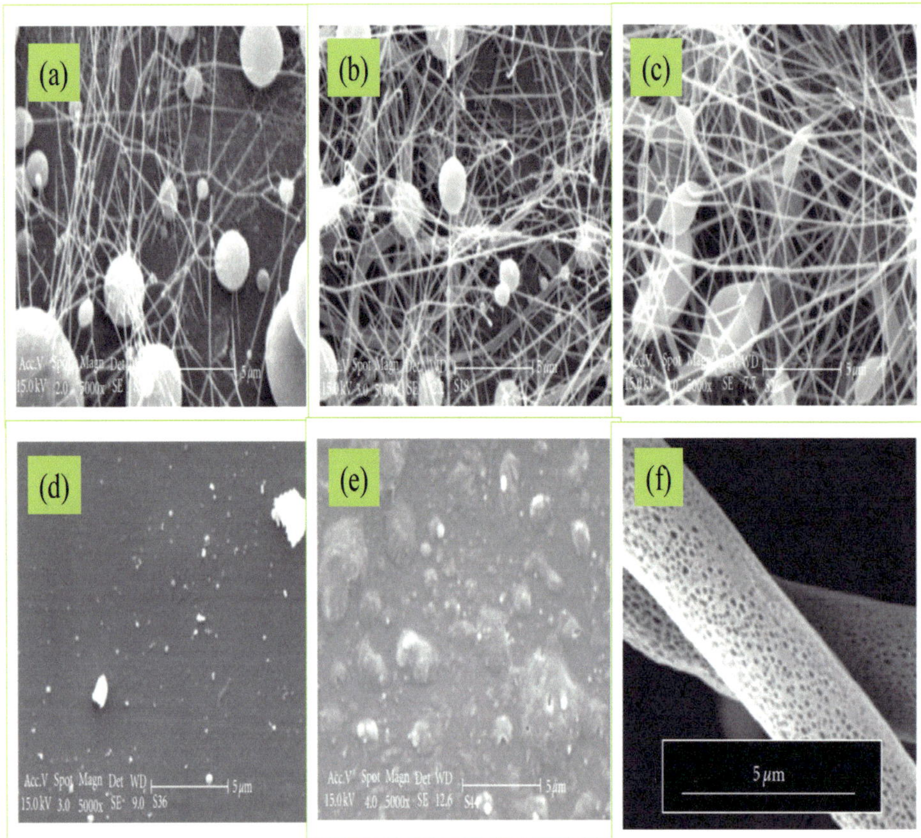

Figure 10.
SEM images of 5% PCL solutions dissolved in different solvents: (a) glacial acetic acid, (b) 90% acetic acid, (c) methylene chloride/ DMF = 4/1, (d) glacial formic acid, (e) and formic acid/acetone, along with (f) SEM images of PVB nanofibers prepared from 10 wt% THF/DMSO (9/1 v/v) [39, 43].

and dimethylformamide (DMF) [29]. The conductivity and dipole moment of the solvent are also very vital. Jarusuwannapoom et al. from the test of 18 solvents concluded only five solvents (ethyl acetate, DMF, THF, methyl ethyl ketone, and 1,2-dichloroethane) were feasible to be used in the electrospinning of PS, since these solvents displayed relatively superior conductivity and dipole moment [42].

4.3 Effects of capillary and collector distance

The distance between the metallic needle tip and collector plays an essential role in determining the morphology of electrospun nanofibers. Similar to the applied electric field, viscosity, and flow rate, the distance between the metallic needle tip and collector also varies with the polymer system. The nanofiber morphology could be easily affected by the distance because it depends on the deposition time, evaporation rate, and whipping or instability interval [44]. Hence, a critical distance needs to be maintained to prepare smooth and uniform electrospun nanofibers, and any changes on either side of the critical distance will affect the morphology of the nanofibers [45]. Numerous research groups have studied the effect of the distance between the needle tip and collector and concluded that defective and large-diameter nanofibers are formed when this distance is kept small, whereas the diameter of the nanofibers decreased as the distance was increased [27, 44, 46]. However, there are cases where no effect on the morphology of the nanofibers was observed with a change in the distance between the metallic needle and collector [47].

5. Electrospray

Electrospray is a technique that uses electricity to disperse a liquid and produce fine aerosol. High voltage is applied to a liquid supply usually a glass or metallic capillary. The liquid on reaching to the tip of the capillary tube (in ideal conditions) forms a Taylor cone, which produces a liquid jet through its top. Varicose waves (twisted and lengthened waves) on the surface of the jet lead to the creation of small and highly charged liquid droplets, which due to Coulomb repulsion are radially dispersed. Electrospray deposition technique is very famous among chemical and medical researchers. This system has several advantages (such as high drug-loading efficiency and self-dispersion) over conventional methods.

5.1 Evolution history of electrospray

In the late sixteenth century, Gilbert [48] described the behavior of magnetic and electrostatic phenomena. He noticed that, in the presence of a charged piece of amber, a drop of water deformed into a cone. This effect was clearly related to electrospray and considered to be the first such observation. Gilbert did not record/elaborate the observation. In 1750, a French clergyman and physicist Jean-Antoine (Abbé) Nollet noticed that water flowing from a vessel aerosolized when the vessel was electrified and placed near electrical ground. He also observed that if a person is electrified due to a connection to a high-voltage generator, he would not bleed normally if he were to cut himself. The blood would only spray from the wound [49]. In 1882, Rayleigh theoretically estimated the maximum amount of charge a liquid droplet could carry [50]. This theory is now known as the "Rayleigh limit." He predicted that a droplet on reaching Rayleigh limit would move as fine jets of liquid. More than 100 years later, Rayleigh limit theory was confirmed experimentally [51]. In 1914, Zeleny studied the behavior of fluid droplets at the end of glass capillaries. The work was later published [52]. The report presented experimental evidence for several electrospray-operating regimes (dripping, burst, pulsating, and cone jet). A few years later, the first time-lapse images of the dynamic liquid meniscus were captured [53]. Between 1964 and 1969, Taylor produced the theoretical groundwork of electrospraying [54–56]. Taylor demonstrated a cone formation by the fluid droplet when an electric field was applied. This characteristic droplet shape is named as the Taylor cone. He further worked with Melcher to develop the "leaky dielectric model" for conducting fluids [57].

6. Mechanism of electrospray

Electrospraying is a technique (**Figure 11**) used for liquid atomization that achieves the stretching and breakup of polymeric solution via electrical forces to obtain micro- or nanoscale particles. Many forces control and direct the electrospraying process; the most important are (i) gravity of polymeric solution, (ii) electrostatic force generated (from external electric field) between nozzle and collector, (iii) repulsion force (Coulomb forces) among adjacent charged carriers on the surface of jet, (iv) viscoelastic force of polymeric solution, (v) interfacial surface tension between air and liquid, and (vi) frictional force between the charged jet and the surrounding air. Among these forces, electrostatic, repulsion, viscoelastic forces, and surface tension affect the stretching and atomization of the droplets during the electrospraying process. When a solution flows out of the nozzle (needle), the charge distribution and carried charge quantity on the surface

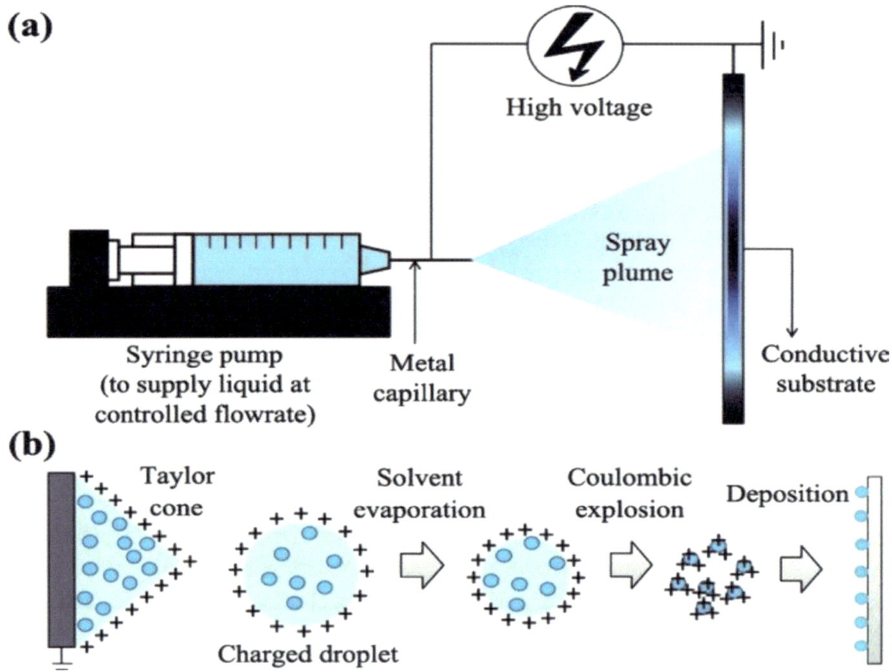

Figure 11.
(a) Basic electrospraying setup and (b) electrospraying mechanism [59].

of the solution will change in varying degrees, according to its electrical conductivity and dielectric constant, because of the polarization effect coming from the external electric field. At the same time, initially, the uncharged liquid becomes charged jet and is further stretched toward the direction of electrostatic attraction. However, compared to gravity and electrostatic force that accelerates the movement and stretching of polymeric solution from the nozzle to the collector, the surface tension and viscoelastic ones prevent this movement and elongation because of their opposite behavior on the electrosprayed solution. When these forces reach a balance at a certain range, the droplets at the tip of the nozzle are stretched from the spherical surface into conical surface. In 1964, Taylor proposed, for a perfectly conducting liquid, a first explanation of the conical shape formation, corresponding to a hydrostatic balance between electrostatic forces and surface tension. The presence of the conical surface at the tip of the nozzle during electrospraying is also called Taylor cone. According to Rayleigh's theory, when the charge quantity distributed on the surface of droplets reaches the value between 50 and 80% of the Rayleigh limit, the breakup and fission of charged droplets occur due to Coulomb repulsions among charged droplets [58]. A classical electrospray setup is considered, with the glass capillary tube situated at a distance (d) from a grounded counter-electrode. The liquid being sprayed is characterized by its viscosity (μ), surface tension (γ), conductivity (κ), and relative permittivity (ε_r).

6.1 Effect of small electric fields on liquid menisci

The liquid meniscus adopts a semispherical shape at the needle tip under the influence of surface tension. Application of the voltage (V) will bring into effect the electric field [60]:

$$E = \frac{2V}{r \ln(4d/r)}$$

where (r) is the liquid radius of curvature. The electric field causes the polarization of the liquid into negative and positive charges. At low voltage, the liquid assumes an equilibrium geometry with a small radius of curvature.

6.2 Taylor cone

A Taylor cone is formed at the applied electric field above the threshold value. Theoretically the shape of the cone depends on two assumptions: first the cone has equipotential surface, and second the cone is in steady-state equilibrium [54]. To meet both of these conditions, the electric field must have ($R^{1/2}$) dependence and azimuthal symmetry to counter the surface tension. The answer to this problem is given as below:

$$V = V_0 + AR^{1/2}P_{1/2}(\cos\theta_0)$$

where (($V=V_o$) equipotential surface) occurs at a value of (θ_o) (regardless of R) leading to an equipotential cone. The angle necessary for ($V=V_o$) for all R is a zero of the Legendre polynomial of order 1/2, P1/2 ($\cos\theta_o$). There is only one zero between 0 and Π at 130.7099°, which is the complement of the Taylor's now famous 49.3° angle.

6.3 Singularity development

The top of the conical meniscus cannot turn into substantially small. A singularity develops when the hydrodynamic relaxation time $\tau_H = \frac{\mu r}{\gamma}$ becomes larger than the charge relaxation time $\tau_C = \frac{\epsilon_r \epsilon_0}{\kappa}$ [61]. The undefined symbols stand for characteristic length (r) and vacuum permittivity (ε_0). Due to intrinsic varicose unsteadiness, the charged jet that eject from the cone ruptures into small charged droplets, which are dispersed radially by the space charge.

6.4 Closing the electrical circuit

The charged liquid ejects through the cone top and is collected as charged droplets on the collector which act as electrode. To balance the loss charge, the

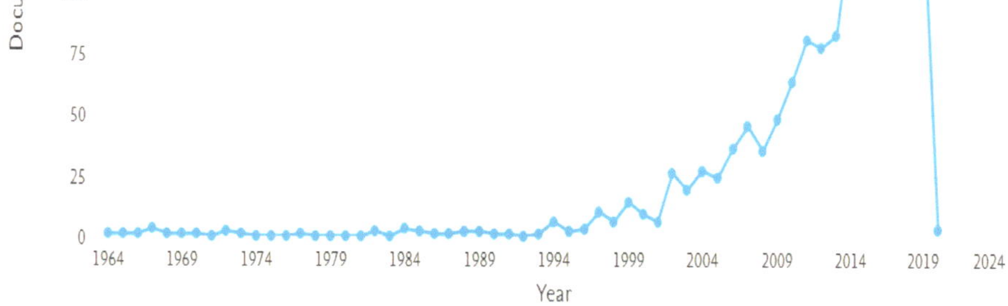

Figure 12.
Scopus data showing year-wise publication in the area of electrospraying. The data is taken on February 09, 2019.

negative charge (which is in excess) is neutralized electrochemically at the nozzle tip. A disproportion between the amount of charge produced electrochemically and that of the lost at the cone top could result in numerous electrospray regimes. For a stable electrosprays, the potential at interface (metal/liquid) self-regulates to produce the equal amount of charges that are lost through the cone tip [62].

Various institution and disciplines around the world are working on electrospray technique to prepare nano- and microparticles for a number of applications ranging from food to biomedical. **Figures 12–15** show the research interest in the area of electrospraying. However, even with the widespread use of the electrospraying technique, the understanding of this method is still very limited.

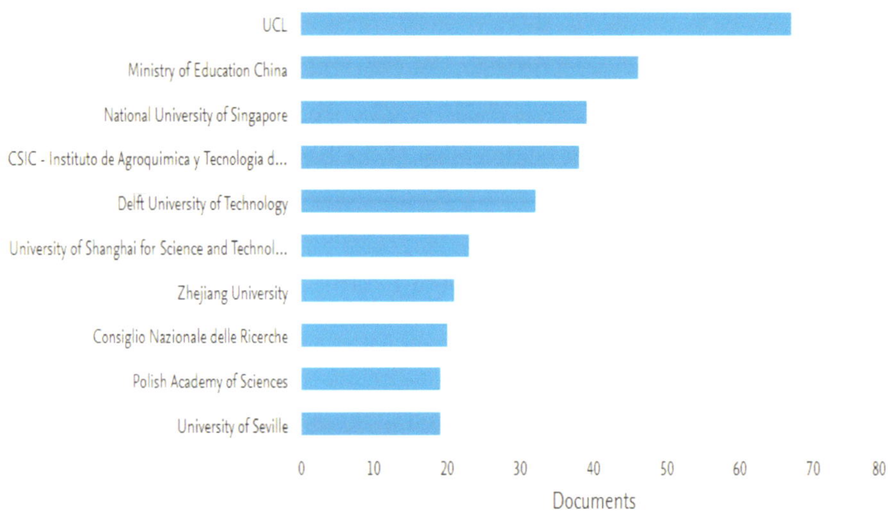

Figure 13.
Country-wise Scopus data of electrospraying. The data is taken on February 09, 2019.

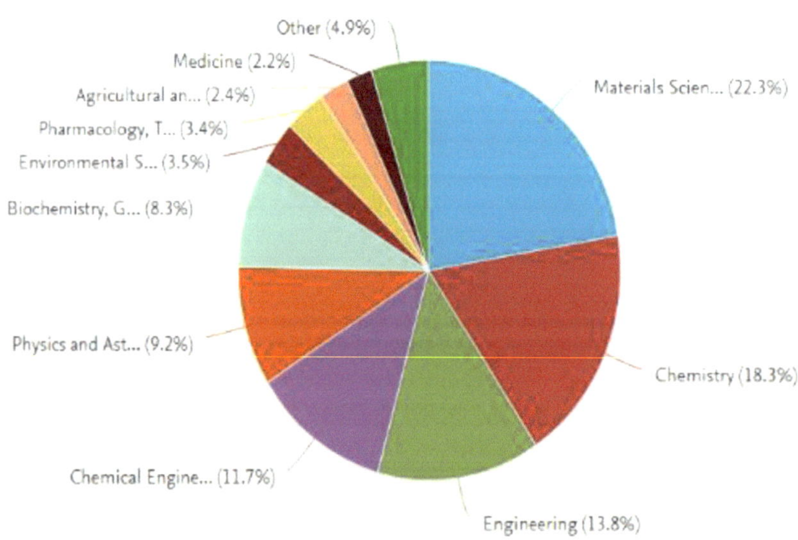

Figure 14.
Showing Scopus data subject wise where electrospraying has been used. The data is taken on February 09, 2019.

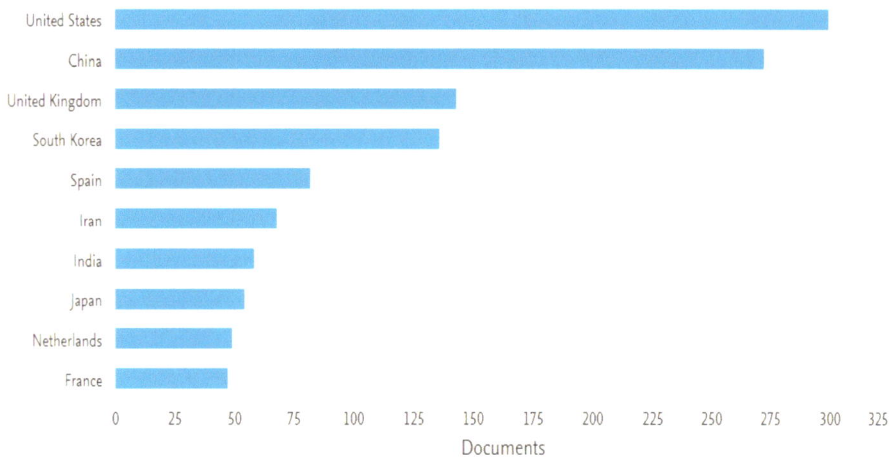

Figure 15.
Showing Scopus data of the institutions, which are working in area of electrospraying. The data is taken on February 09, 2019.

7. Effects of parameters on electrospraying

Similar to electrospinning, there are various factors that affect the electrospraying process. These factors are classified as operating and solution parameters. The operating parameters include (i) applied voltage, (ii) working distance, (iii) liquid flow rate, and (iv) nozzle type. The solution parameters include (i) electrical conductivity, (ii) surface tension, (iii) viscosity, (iv) vapor pressure, and (v) dielectric constant. All these parameters directly affect the formation of the Taylor cone, which leads to the production of particles having a low mean diameter and narrow particle size distribution. Therefore, to gain a better understanding of the electrospraying technique and achieve low mean diameter and narrow particle size distribution, it is essential to thoroughly understand the effects of all of these governing parameters [58].

7.1 Applied voltage

Applied voltage is one of the essential parameters, which needs to be optimized to get low diameter and narrow particle distribution. Voltage applied between the nozzle and collector not only influences the stretching of the jet but also the formation of the Taylor cone at the end of the nozzle. Increase in the applied voltage increases the electrostatic forces that are acting on the surface of the charged droplets. This causes the spray mode to gradually take its route from dripping to multi-jet and then to a stable cone jet (**Figure 16** and **Table 1**). When the electrostatic forces acting on the surface of the spray droplets are not sufficiently strong to overcome surface and viscoelastic tensions (low applied voltage causes drop-to-drop mode), film formation occurs on the collector. In the dripping area, electrified liquid form drops at the capillary end till the gravitational and electrical forces overcome their surface tension. The droplets emerge at the end of the capillary at low frequency and maintain spherical morphologies due to gravitational force and surface tension. As the voltage is increased, the shape of the ejected droplet is affected by its wetting properties, which allows the formation of particles with high average diameter and wide size distribution. At higher voltages, the jet mode changes to stable conical jet mode (**Figure 17**). At this point the electrostatic forces

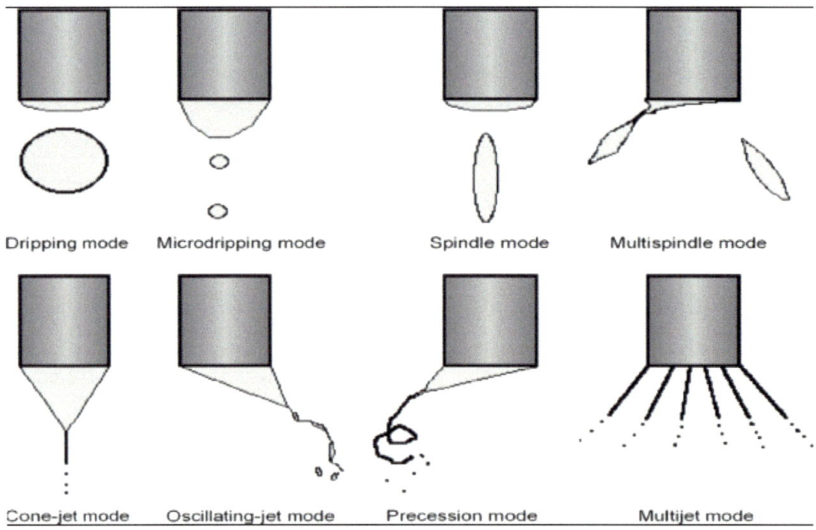

Figure 16.
Schematic showing various possible jet modes during electrospraying process adopted from [63].

Spraying mode types			
Dripping		Jet	
Dripping (A)	Regular large droplets.	Cone jet (E)	A smooth and stable liquid jet.
Micro-dripping (B)	Fine droplets with a narrow size distribution and low frequency, generally occurs at low liquid flow rates.	Precession (F)	The liquid jet rotates around the capillary axis.
Spindle (C)	Occurs at high flow rates along with increased electrical forces and produces elongated spindles.	Oscillating (G)	Liquid jet oscillates in its own plane.
Multi-spindle (D)	Occurs at higher flow rates along with increased electrical forces and produces multiple elongated spindles.	Multi-jet (H)	A few fine jets on the circumference of the capillary exit.

Table 1.
A summary of the main spraying jet modes (adopted from [63]).

are strong enough to overcome surface and viscoelastic tension, which leads to a balance between different forces and fabrication of low mean diameter particles of narrow size distribution [63, 64].

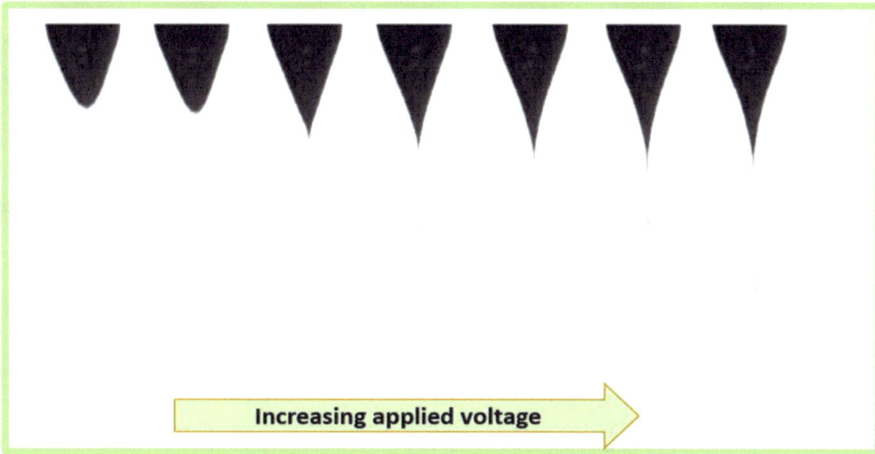

Figure 17.
Showing the switching of a liquid meniscus (subjected to increasing applied voltage) into a conical shape and a stable jet, adopted from [62, 65].

7.2 Flow rate

The flow rate of the electrospray solution is also an essential factor in the formation of the Taylor cone. The optimization of this parameter depends on two intrinsic characteristics of the solutions: (i) vapor pressure of the solvent (ii) and electrical conductivity of the solutions. A stable Taylor cone prefers low flow rates because the uncharged drop at the tip of the nozzle needs sufficient time polarization. This polarization generates charges on the droplet surface, which in turn promote Taylor cone formation. At higher flows, the decrease in the polarization/partial polarization of the droplets result in continuous dripping of droplets due to gravity. Having said this, for different materials the flow rate to form a stable jet varies. The use of a solvent with a high-saturation vapor pressure needs a higher flow rate to avoid nozzle blocking, when the solvent evaporates. A lower flow rate is recommended for solutions prepared from solvents having low saturated vapor pressure. This is done to ensure complete removal (evaporation) of the solvent during solidification process. Another factor that might affect the optimization of flow rate is the conductivity of the solutions. It is a known fact that conductivity of a solution influences the droplet polarization time. This means, it will also influence the adjustment of an adequate flow rate. Since low conductivity solution will need more time to polarize and generate charges on the droplet, hence in such case low flow rate is recommended. Reverse is the case with high-conductivity solutions [66].

7.3 Working distance

The working distance between the metallic nozzle tip and collector also plays an essential role in determining the formation of a table Taylor cone. Similar to the applied electric field, viscosity, and flow rate, the distance between the metallic needle tip and collector also varies for different polymer systems. Smeets et al. achieved the formation of a stable Taylor cone using a lager tip-to-collector distance and low flow rate. The working distance between the tip and the collector is controlled to achieve a critical electric field (at constant applied field) and ensure the formation of the Taylor cone. At shorter distance, the electric field is higher. This leads to the formation of small particles. However, this is only possible when the solvent is more volatile; otherwise swelled particles are collected on the collector. Long working distance may cause lower yield. This is attributed mainly to the loss of material to the surrounding due to the turbulent droplet flight. Having said this, long working distance is needed when denser polymer particles are required. The particles do not swell on the collectors because of the availability of sufficient time for the droplet's solvent to evaporate in the flight from nozzle tip to the collector. For low conductive solution with more volatility, nozzle to collector distance is smaller and vice versa for the high-conductivity and low-volatility solution [64].

7.4 Electrical conductivity

In the process of electrospraying, a solution is stretched and converted into a jet at the tip of the nozzle due to the electrostatic force, when an electric field is applied between the nozzle tip and collector. A balance between gravity, electrostatic force, Coulomb repulsions, viscoelastic and frictional forces, and surface tension leads to the formation of a Taylor cone at the tip of the nozzle (**Figures 11** and **17**). The strength of the electrostatic force and Coulomb repulsions that are acting on the electrosprayed droplets depends on the amount of polarization charge on the droplets, which in turn depends on the strength of the applied electric field. These electrostatic attractions counteract the surface tension and viscoelastic forces, which then leads to the formation of Taylor cone. When the electrical conductivity of the solution is too high, this high conductivity causes increased Coulomb repulsions among charged droplets and breakup of the charged droplets. This phenomenon leads to the modification of the balance between gravity, electrostatic force, Coulomb repulsions, viscoelastic and frictional forces, and surface tension and a change from a stable cone jet to an unstable jet. Similarly, a too low electrical conductivity causes decreased electrostatic attractions on the droplets, which leads to the same unbalance among the abovementioned forces and results neither in the Taylor cone formation at the nozzle tip nor the formation of particle. Hence, a critical conductivity of the solution is needed, which results in the formation of a stable Taylor cone. Jaworek and Xie et al. suggested that a electrohydrodynamic process is carried out under cone-jet mode, when the electrical conductivity of the solution is in the range of 10^{-11} to 10^{-1} S/m [67].

7.5 Surface tension and viscosity

As discussed earlier, unbalance between the forces affects the formation of a Taylor cone at the tip of the nozzle during the electrospraying. One of these forces is surface tension. Cloupeau and Smith suggested that the surface tension value of the solution used in electrospraying must not exceed 50 mN/m, though some studies reported more value for glycerine (63 mN/m) and water (72.8 mN/m) [68–71]. A high surface tension of a solution will result in a corona discharge at the tip of the nozzle, which will cause a stable cone jet to change into an irregular spraying and an asymmetrical mode. Similar to surface tension, viscosity of the solutions also affects the formation of a stable Taylor cone. Since the direction of the viscoelastic force is opposite to the direction of the gravity and electrostatic attraction, when the viscosity of the solution is low, the combined viscoelastic force and surface tension are sufficiently strong to counteract the gravity and electrostatic attraction forces. This leads to dripping instead of cone jet. A high-viscosity solution does not allow the formation of a stable Taylor cone due to the drying of the polymeric particles, which block the tip of the capillary and therefore restricting the formation of a stable Taylor cone. Thus, in order to achieve a stable con jet during electrospraying process, a critical range of viscosity is needed. This range will be different for different for polymeric systems. For some the range may be from 1.5 to 5500 mPa·s [52, 53].

7.6 Vapor pressure and dielectric constant

Boiling point of the solution in electrospraying also affects the formation of Taylor cone. Since the evaporation of solvent and the formation of Taylor cone at the tip of the nozzle occur simultaneously in jetting, thus a too low boiling point (high vapor pressure) changes the shape of the stable cone jet into an unstable one. Furthermore, a low boiling point of solution will result in the drying of the polymeric particles at the tip of the nozzle and will block the tip and stop the process. The effect of the dielectric constant on the cone-jet formation is similar to the electrical conductivity. A very high and too low dielectric constant is not favorable to obtain a stable cone jet during electrospraying process (e.g., polycaprolactone (PCL)). Some research published on the electrospinning of PS correlated to the dipole moment and the dielectric constant of the solutions. To date, very few articles have studied the effect of dielectric constant on the outcomes of electrospraying; much work is needed to completely understand its effects [42, 72].

8. Conclusion

Electrospinning is a technique used for the fabrication of nanofibers, whereas electrospray is a technique used for the preparation of micro- or nanoscale size, mono-dispersed particles. Both of these electrodynamic techniques have been receiving increasing attention not only in the scientific community but also in the food, environmental, and biomedical industries. The main apparatus used for both of these processes is almost the same. Both need electric voltage to induce charges on the droplet, which at optimized electric filed leads to micro- or nanofibers and micro- or nanoparticles. Rayleigh in 1882, for the first time, theoretically estimated the maximum amount of charge a liquid droplet could carry to change in a jet. This theory is now known as the "Rayleigh limit." His predicted that a droplet on reaching Rayleigh limit would move as fine jets of liquid. More than 100 years later, Rayleigh limit theory was confirmed experimentally. Beside electric filed there are other operating and solution parameters that need to be optimized before we obtain our desired products. The particle and fiber formation mechanisms and principles of electrospraying and electrospinning are still complicated, and more studies are needed understand both of these processes.

Author details

Sajjad Haider[1*†], Adnan Haider[2†], Abdulaziz A. Alghyamah[1†], Rawaiz Khan[3†], Waheed A. Almasry[1†] and Naeem Khan[3†]

1 Department of Chemical Engineering, College of Engineering, King Saud University, Riyadh, KSA, Saudi Arabia

2 Department of Chemistry, Kohat University of Science and Technology, Kohat, KP, Pakistan

3 Engineer Abdullah Bugshan Research Chair for Dental and Oral Rehabilitation, College of Dentistry, King Saud University, Riyadh, Saudi Arabia

*Address all correspondence to: shaider@ksu.edu.sa

†All authors have equal contribution.

References

[1] Ohkawa K, Cha D, Kim H, Nishida A,Yamamoto H.Electrospinning of chitosan. Macromolecular Rapid Communications. 2004;**25**:1600-1605

[2] Buchko CJ, Chen LC, Shen Y, Martin DC. Processing and microstructural characterization of porous biocompatible protein polymer thin films. Polymer. 1999;**40**:7397-7407

[3] Saeed K, Haider S, Oh TJ, Park SY. Preparation of amidoxime-modified polyacrylonitrile (PAN-oxime) nanofibers and their applications to metal ions adsorption. Journal of Membrane Science. 2008;**322**:400-405

[4] Morton WJ. Method of dispersing fluids. US Patent 705691; 1902

[5] Cooley JF. Apparatus for electrically dispersing fluids. US Patent 692631; 1902

[6] Formhals A. Process and apparatus for preparing artificial threads. US Patent 1975504; 1934

[7] Doshi J, Reneker DHJ. Electrospinning process and applications of electrospun fibers. Journal of Electrostatics. 1995;**35**:151-160

[8] Lim SK, Lee S-K, Hwang S-H, Kim H. Photocatalytic deposition of silver nanoparticles onto organic/inorganic composite nanofibers. Macromolecular Materials and Engineering. 2006;**291**:1265-1270

[9] Peng Y, Dong Y, Fan H, Chen P, Li Z, Jiang Q. Preparation of polysulfone membranes via vapor-induced phase separation and simulation of direct-contact membrane distillation by measuring hydrophobic layer thickness. Desalination. 2013;**316**:53-66

[10] Yongquan D, Ming W, Lin C, Mingjun L. Preparation, characterization of P(VDF-HFP)/[bmim]BF4 ionic liquids hybrid membranes and their pervaporation performance for ethyl acetate recovery from water. Desalination. 2012;**295**:53-60

[11] Chen L, Bromberg L, Hatton TA, Rutledge GC. Catalytic hydrolysis of p-nitrophenyl acetate by electrospun polyacrylamidoxime nanofibers. Polymer. 2007;**48**:4675-4682

[12] Kijeńska E, Prabhakaran MP, Swieszkowski W, Kurzydlowski KJ, Ramakrishna S. Electrospun bio-composite P(LLA-CL)/collagen I/collagen III scaffolds for nerve tissue engineering. Journal of Biomedical Materials Research Part B: Applied Biomaterials. 2012;**100B**:1093-1102

[13] Katepalli H, Bikshapathi M, Sharma CS, Verma N, Sharma A. Synthesis of hierarchical fabrics by electrospinning of PAN nanofibers on activated carbon microfibers for environmental remediation applications. Chemical Engineering Journal. 2011;**171**:1194-1200

[14] Subramanian S, Seeram R. New directions in nanofiltration applications—Are nanofibers the right materials as membranes in desalination? Desalination. 2013;**308**:198-208

[15] Feng C, Khulbe KC, Matsuura T, Tabe S, Ismail AF. Preparation and characterization of electro-spun nanofiber membranes and their possible applications in water treatment. Separation and Purification Technology. 2013;**102**:118-135

[16] Huang Z-M, Zhang YZ, Kotaki M, Ramakrishna S. A review on polymer nanofibers by electrospinning and their applications in nanocomposites. Composites Science and Technology. 2003;**63**:2223-2253

[17] Wang X, Drew C, Lee SH, Senecal KJ, Kumar J, Samuelson LA. Electrospun nanofibrous membranes for highly sensitive optical sensors. Nano Letters. 2002;**2**:1273-1275

[18] Hohman MM, Shin M, Rutledge G, Brenner MP. Electrospinning and electrically forced Jets. II. applications. Physics of Fluids. 2001;**13**:2221-2236

[19] Bognitzki M, Czado W, Frese T, Schaper A, Hellwig M, Steinhart M, et al. Nanostructured Fibers via Electrospinning. Advanced Materials. 2001;**13**:70-72

[20] Hou H, Jun Z, Reuning A, Schaper A, Wendorff JH, Greiner A. Poly(p-xylylene) nanotubes by coating and removal of ultrathin polymer template fibers. Macromolecules. 2002;**35**:2429-2431

[21] Wei G, Ma PX. Nanostructured biomaterials for regeneration. Advanced Functional Materials. 2008;**18**:3566-3582

[22] Ramakrishna S, Fujihara K, Teo WE, Yong T, Ma Z, Ramaseshan R. Electrospun nanofibers: Solving global issues. Materials Today. 2006;**9**:40-50

[23] Haider S, Al-Zeghayer Y, Ahmed Ali F, Haider A, Mahmood A, Al Masry W, et al. Highly aligned narrow diameter chitosan electrospun nanofibers. Journal of Polymer Research. 2013;**20**:1-11

[24] Bae H-S, Haider A, Selim KMK, Kang D-Y, Kim E-J, Kang I-K. Fabrication of highly porous PMMA electrospun fibers and their application in the removal of phenol and iodine. Journal of Polymer Research. 2013;**20**:1-7

[25] Sill TJ, von Recum HA. Electrospinning: Applications in drug delivery and tissue engineering. Biomaterials. 2008;**29**:1989-2006

[26] Deitzel JM, Kleinmeyer J, Harris D, Beck Tan NC. The effect of processing variables on the morphology of electrospun nanofibers and textiles. Polymer. 2001;**42**:261-272

[27] Baumgarten PK. Electrostatic spinning of acrylic microfibers. Journal of Colloid and Interface Science. 1971;**36**:71-79

[28] Laudenslager MJ, Sigmund WM. Encyclopedia of Nanotechnology. Bhushan B. editor. Berlin, Germany: Springer. 2012. pp. 769-775

[29] Megelski S, Stephens JS, Bruce Chase D, Rabolt JF. Micro- and nanostructured surface morphology on electrospun polymer fibers. Macromolecules. 2002;**35**:8456-8466

[30] Zeleny J. The role of surface instability in electrical discharges from drops of alcohol and water in air at atmospheric pressure. Journal of the Franklin Institute. 1935;**219**:659-675

[31] Shamim Z, Saeed B, Amir T, Saied RA, Rogheih D. The effect of flow rate on morphology and deposition area of electrospun nylon 6 nanofiber. Journal of Engineered Fibers and Fabrics. 2012;**7**:42-49

[32] Zong X, Kim K, Fang D, Ran S, Hsiao BS, Chu B. Structure and process relationship of electrospun bioabsorbable nanofiber membranes. Polymer. 2002;**43**:4403-4412

[33] Theron S, Zussman E, Yarin A. Experimental investigation of the governing parameters in the electrospinning of polymer solutions. Polymer. 2004;**45**:2017-2030

[34] Reneker DH, Kataphinan W, Theron A, Zussman E, Yarin AL. Nanofiber garlands of polycaprolactone by electrospinning. Polymer. 2002;**43**:6785-6794

[35] Pillay V, Dott C, Choonara YE, Tyagi C, Tomar L, Kumar P, et al. A review of the effect of processing variables on the fabrication of electrospun nanofibers for drug delivery application. Journal of Nanomaterials. 2013;**2013**:1-22

[36] Zander NE. Hierarchically structured electrospun fibers. Polymers. 2013;**5**:19-44

[37] Luzio A, Canesi E, Bertarelli C, Caironi M. Electrospun polymer fibers for electronic applications. Materials. 2014;**7**:906-947

[38] Lannutti J, Reneker D, Ma T, Tomasko D, Farson D. Electrospinning for tissue engineering scaffolds. Materials Science and Engineering: C. 2007;**27**:504-509

[39] Kanani AG, Bahrami SH. Effect of changing solvents on poly(-caprolactone) nanofibrous webs morphology. Journal of Nanomaterials. 2011;**2011**:1-10

[40] Fong H, Chun I, Reneker DH. Beaded nanofibers formed during electrospinning. Polymer. 1999;**40**:4585-4592

[41] Zhang Y, Feng Y, Huang Z, Ramakrishna S, Lim CT. Fabrication of porous electrospun nanofibers. Nanotechnology. 2006;**17**:901-908

[42] Jarusuwannapoom T, Hongrojjanawiwat W, Jitjaicham S, Wannatong L, Nithitanakul M, Pattamaprom C, et al. Effect of solvents on electro-spinnability of polystyrene solutions and morphological appearance of resulting electrospun polystyrene fibers. European Polymer Journal. 2005;**41**:409-421

[43] Lubasova D, Martinova L. Controlled morphology of porous polyvinyl butyral nanofibers. Journal of Nanomaterials. 2011;**2011**:1-6

[44] Matabola KP, Moutloali RM. The influence of electrospinning parameters on the morphology and diameter of poly(vinyledene fluoride) nanofibers-effect of sodium chloride. Journal of Materials Science. 2013;**48**:5475-5482

[45] Bhardwaj N, Kundu SC. Electrospinning: A fascinating fiber fabrication technique. Biotechnology Advances. 2010;**28**:325-347

[46] Wang T, Kumar S. Electrospinning of polyacrylonitrile nanofibers. Journal of Applied Polymer Science. 2006;**102**:1023-1029

[47] Zhang C, Yuan X, Wu L, Han Y, Sheng J. Study on morphology of electrospun poly(vinyl alcohol) mats. European Polymer Journal. 2005;**41**:423-432

[48] Gilbert W. De Magnete, Magneticisque Corporibus, et de Magno Magnete Ellure (On the Magnet and Magnetic Bodies, and on That Great Magnet the Earth). London: Peter Short; 1628

[49] Grimm RL. 2. Fundamental Studies of the Mechanisms and Applications of Field-Induced Droplet Ionization Mass Spectrometry and Electrospray Mass Spectrometry. Dissertation (Ph.D.), California Institute of Technology. Available from: https://resolver.caltech.edu/CaltechETD:etd-10092005-222651

[50] Rayleigh L. XX. On the equilibrium of liquid conducting masses charged with electricity. The London, Edinburgh, and Dublin Philosophical Magazine and Journal of Science. 1882;**14**(1):184-186

[51] Gomez A, Tang K. Charge and fission of droplets in electrostatic sprays. Physics of Fluids. 1994;**6**(1):404-414

[52] Zeleny J. The Electrical discharge from liquid points, and a hydrostatic method of measuring the electric

Intensity at their surfaces. Physical Review. 1914;**3**(2):69-91

[53] Zeleny J. Instability of electrified liquid surfaces. Physical Review. 1917;**10**(1):1-6

[54] Taylor G. Disintegration of water droplets in an electric field. Proceedings of the Royal Society A. 1964;**280**(1382):383-397

[55] Taylor G. The force exerted by an electric field on a long cylindrical conductor. Proceedings of the Royal Society of London, A: Mathematical, Physical and Engineering Sciences. 1965;**291**(1425):145-158

[56] Taylor GI, Van Dyke MD. Electrically driven jets. Proceedings of the Royal Society of London A: Mathematical, Physical and Engineering Sciences. 1969;**313**(1515):453-475

[57] Melcher JR, Taylor G. Electrohydrodynamics. Annual Review of Fluid Mechanics. 1969;**1**:111-146

[58] Zhang S, Campagne C, Salaün F. Applied Sciences. 2019;**9**:402

[59] Saallah S, Lenggoro W. Influence of solvent selection in the electrospraying process of polycaprolactone. Kona Powder and Particle Journal. 2018;**402**:1-36

[60] Loeb LB, Kip AF, Hudson GG, Bennett WH. Pulses in negative point-to-plane corona. Physical Review. 1941;**60**(10):714-722

[61] Fernández de la Mora J, Loscertales IG. The current emitted by highly conducting taylor cones. Journal of Fluid Mechanics. 1994;**260**:155-184

[62] Van Berkel GJ, Zhou FM. Characterization of an electrospray ion source as a controlled-current electrolytic cell. Analytical Chemistry. 1995;**67**(17):2916-2923

[63] Smith A. Electrohydrodynamic atomization produced nanoparticles for the targeted delivery of cancer chemotherapeutics [Doctor of Philosophy thesis]. England: University of Portsmouth; 2015

[64] Smeets A, Clasen C, Van den Mooter G. Electrospraying of polymer solutions: Study of formulation and process parameters. European Journal of Pharmaceutics and Biopharmaceutics. 2017;**119**:114-124

[65] Yurteri CU, Hartman RPA, Marijnissen JCM. Producing pharmaceutical particles via electrospraying with an emphasis on nano and nano structured particles - A review. Kona Powder and Particle Journal. 2010;**28**:91-115

[66] Jayasinghe SN, Nicholson T. Stable electric-field driven cone-jetting of concentrated biosuspensions. Lab on a Chip. 2006;**6**:1086-1090

[67] Jaworek A. Micro- and nanoparticle production by electrospraying. Powder Technology. 2007;**176**:18-35

[68] Xie J, Jiang J, Davoodi P, Srinivasan MP, Wang C-H. Electrohydrodynamic atomization: A two-decade effort to produce and process micro-/nanoparticulate materials. Chemical Engineering Science. 2015;**125**:32-57

[69] Cloupeau M. Recipes for use of EHD spraying in cone-jet mode and notes on corona discharge effects. Journal of Aerosol Science. 1994;**25**:1143-1157

[70] Smith DPH. The electrohydro-dynamic atomization of liquids. IEEE Transactions on Industry Applications. 1986;**IA-22**:527-535

[71] Cloupeau M, Prunet-Foch B. Electrohydrodynamic spraying functioning modes: A critical

review. Journal of Aerosol Science. 1994;**25**:1021-1036

[72] Pham QP, Sharma U, Mikos AG. Electrospinning of Polymeric Nanofibers for Tissue Engineering Applications: A review. Tissue Engineering. 2006;**12**:1197-1211

Section 2

Electrospinning

Chapter 2

Electrospinning and Drug Delivery

Marilena Vlachou, Angeliki Siamidi and Sotiria Kyriakou

Abstract

A detailed account of the construction, properties, and practical applications of electrospinning for the fabrication of high-quality ultrafine fibers, suitable for drug delivery, is given. With respect to the electrospinning method, various parameters are of crucial importance. The electrospinning parameters are classified as solution properties, process parameters, and environmental conditions. The solution properties include the polymer concentration, molecular weight and viscosity, the solution conductivity and relative volatility, volatility of the solvent, surface tension, and dielectric constant. The process parameters refer to the flow rate, the applied voltage, the needle diameter, and the distance between the tip of the needle and collector and the geometry of the collector. The environmental conditions include the relative humidity and temperature. All these factors are responsible for a flawless electrospinning process, which leads to the formation of the desirable electrospun nanofibers with the requisite characteristics. In this chapter, it has been shown that the electrospinning technology could provide a useful method for modifying drug release behavior and opens new routes for the development of effective and tailor-made drug release carriers.

Keywords: nanofibers, electrospinning, electrospinning parameters, polymers, oral drug delivery

1. Introduction

During the last years, nanofibers have become increasingly attractive as drug delivery systems, mainly because they enhance the delivery of limited absorption drugs by improving the dissolution rates and solubility of drug molecules. Moreover, nanofibrous approaches in preparing stable amorphous drug formulations are extensively profitable [1].

The principal methods used for the fabrication of polymer nanofibers include drawing, template synthesis, phase separation, self-assembly, solvent casting, and electrospinning. However, the latter has become the most frequently used technique because of its ability to afford nanofibers with unique characteristics. These include a very high surface area to volume ratio, a high porosity with a small pore size, improved mechanical properties, degradability, and flexibility in surface functionalities/motifs. All the other fabrication methods have limitations, with respect to the materials used, and moreover, they are laborious and complex processes, resulting in problematic scale-up. Furthermore, compared with the other processing techniques, electrospinning is a simple, user-friendly, reproducible, and continuous process [1–4], and upon the appropriate selection of the electrospinning apparatus and materials, diverse types of fibers, such as core-sheath, porous or hollow structured nanofibers can be produced.

2. Electrospinning

As already mentioned, electrospinning is a simple, highly versatile and robust technique for the production of polymers and a wide range of materials, including ceramic, metallic, and long fibers with diameters from submicron down to nanometer scale. These fibers are produced by feeding a polymer solution, dispersion or melt in a high electric field. It is worth mentioning that the use of melt in the electrospinning process is costly and leads to more difficult production than using polymer solution [1–6].

2.1 Electrospinning equipment

The main setup of an electrospinning equipment involves three main parts all enclosed within a chamber. A typical electrospinning apparatus is shown in **Figure 1**. It is composed of an electrical supply, for generating a high-voltage power supply, a piece of feeding equipment, which consists of a glass syringe with metallic needle filled with the polymer solution, a pump suitable for controlling the flow rate of the polymer solution, and a grounded collector usually made from aluminum foil. The power supply is used to apply tens of kilovolt to the needle, which works as a spinneret, while the pump extrudes the polymer from the syringe to the collector, which can be either rotatable or static [2, 4, 5, 7–9].

2.2 Electrospinning process and methods

2.2.1 Electrospinning process

The working principle of electrospinning is straightforward: at ambient temperature, a polymer solution or melt is ejected from the tip of a needle to a grounded metal collector by applying high voltage between the needle and the collector [2]. In

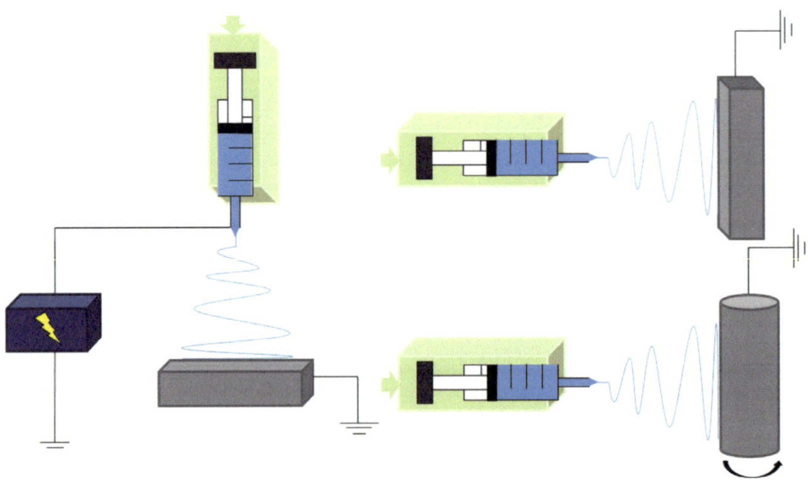

Figure 1.
Typical schematic setup of electrospinning equipment with static (up) and rotatable collector (down).

detail, the electrospinning process starts with the application of high voltage, which creates electric charges that are moving toward the polymer solution in the syringe *via* the metallic needle. The induction of charges on the polymer droplet causes instability within the polymer solution, thus creating an electrically charged jet of polymer solution or melt. Concurrently, a force that opposes the surface tension is produced, by the mutual repulsion of charges, and as a result, the polymer solution flows in the direction of the electric field and is extruded from the needle of the syringe with the aid of a pump [4]. Specifically, the solution jet is ejected from the nozzle of the needle when the voltage exceeds a threshold value because the electric force overcomes the surface tension of the droplet. Each droplet is exposed to a high voltage, and a cone-shaped droplet is formed. This is known as the Taylor cone and is caused due to the electrical voltage, which is the difference voltage between the nozzle and the collector with the counter charge [8]. Subsequently, the charged jet of solution is evaporated or solidified before reaching the metallic collector, where the solid material is collected as a solid interconnected continuous network of small fibers [1, 8, 10]. Regarding the electrospinning process, a stable charge jet can be formed only if the polymer solution has adequate cohesive force. During the process, the internal and external charge forces cause the whipping of the liquid jet, thus permitting the polymer chains to stretch and slide into the solution pushing the jet toward the collector [4]. As a result, the created fibers have enough small diameters to be characterized as diversely functionalized nanofibers because of their surface structure and their potential to modify their morphology and their internal bulk content [1, 4, 11, 12].

2.2.2 Electrospinning methods

The electrospinning technique is very useful for the incorporation of drugs in drug delivery systems. This technique can be reproducible under controlled parameters and is used in many formulations for the creation of new and innovative drug carriers because of their efficiency of transporting the bioactive agents to the target without causing secondary effects in the body. There are different methods for incorporating therapeutic drugs into drug delivery systems with electrospinning, which can greatly influence the properties of the resulting drug-loaded fibrous system. These methods involve blending, coaxial, emulsion, and surface modification electrospinning, which have discrete advantages and disadvantages. According to the physicochemical properties of the drug, the polymeric characteristics and the application of the drug-incorporated fibers, such as the target zone and the required drug release rate, the appropriate method is being selected as not all drugs can be loaded with all of these methods [3, 13, 14].

2.2.2.1 Blending electrospinning

Blending of the therapeutic agent with the appropriate polymeric solution remains the most predominant method for drug loading into nanofibers [3, 13]. This method is simple, compared to others, but some requirements should be met in order to gain the desired results. The polymeric blend improves the mechanical and physicochemical properties equilibrium of the drug-loaded nanofibers and increases effectively the formulation design for drug release, resulting in the manipulation of the release rate by changing the proportion of polymer in the blend [3, 15]. The insufficient solubility of the drug in the polymeric solution, where the drug molecules can shift to a nearby surface of fiber during electrospinning, can trigger the isolate release of the drug into the solution. Thus, the equilibrium among hydrophilic and hydrophobic properties between drugs and polymers is very important during blending electrospinning [3, 7]. The drug release behavior is highly contingent on the distribution of the drug molecule into electrospun nanofibers as well as on the morphology of the nanofibers. In order to achieve perfect encapsulation of the drug inside the electrospun nanofibers, the hydrophobic polyester polymers should interact very well with the hydrophobic or lipophilic drugs, such as rifampicin and paclitaxel, while the hydrophilic polymers, such as gelatin, polyethylene glycol (PEG), and polyvinyl alcohol (PVA), can dissolve hydrophilic drugs, such as doxorubicin. It has been cited that amphiphilic copolymers like the PEG-b-PLA diblock copolymer could significantly enhance drug-loading efficiency and subsequently reduce the burst release of drugs [13]. With the blending electrospinning method, the drug is dissolved or dispersed into the polymer solution to achieve drug encapsulation through a single-step electrospinning, and as a result, fibers are obtained with single phase only [3, 13].

2.2.2.2 Coaxial electrospinning

The coaxial electrospinning method is regarded as one of the most significant breakthroughs, and it is mainly useful for multidrug delivery systems, where the individual drug release behavior is controlled [2, 3, 13]. In this method, there are two liquids inside the spinneret, which minimize the interaction between aqueous-based biological molecules and the organic solvents, in which the polymer is mainly dissolved, and as a result is used for obtaining fibers with core-shell structures [2, 3, 13]. These structures are used in cases where the therapeutic agent is sensible to the environment [3]. Moreover, this method can be used for generating novel structural nanomaterials, such as preparing nanofibers from materials without filament-forming properties enclosing functional liquids within the fiber matrix and encapsulating drugs or biological agents in the core of the polymer nanofibers leading to sustained and controlled drug release [2, 3]. The functionality of biomolecules is improved in coaxial electrospinning because the inner jet is formed by the biomolecule solution, and the outer jet is formed by the polymer solution, which is the co-electrospun. Moreover, the polymeric shell contributes to the sustained and prolonged release of the therapeutic agent as well as protecting the ingredient in the core from direct exposure to the biological environment [3, 13]. In this method, the coaxial fibers have successfully loaded proteins, growth factor, antibiotics, and other biological agents for drug delivery purposes [3, 13]. In coaxial electrospinning, there are a lot of factors, which should be considered in the design step, such as shell and core polymer concentration, molecular weight, and drug concentration [13, 14]. Nevertheless, only a limited portion of the produced fibers can form the proper core/shell structure and this system improves the sustained release of drugs and allows the bioability of unstable biological agents to be maintained [3, 14].

2.2.2.3 Emulsion electrospinning

The emulsion electrospinning method is an important and flexible method for the encapsulation of several drugs into nanofibers as well as a cost-effective and efficient manner for preparing core-shell electrospun nanofibers [3, 14]. In the emulsion electrospinning method, the oil phase is created by the emulsion of the drug or aqueous protein solution in the hydrophobic polymer solution. At the end of the electrospinning, the biomolecule-loaded phase can be distributed within the fibers, if a low molecular weight drug is used, or a core-shell fibrous structure can be configured as macromolecules in the aqueous phase [3, 13, 14]. It has been reported that the ratio of hydrophilic (aqueous) to hydrophobic (polymer) solution is one of the parameters that affect the distribution of the biomolecules within the fibers. Moreover, it plays an important role in regulating the release profile, structural stability, and bioactivity of the encapsulated drug or proteins [3, 13, 14]. It is worth mentioning that the main advantage of emulsion electrospinning against blending electrospinning is the elimination of the need for a common solvent as the drug and the polymer are dissolved in applicable solvents. Numerous hydrophilic drugs and hydrophobic polymeric combinations can be used while maintaining minimal drug contact with the organic solvent during the procedure [3, 13, 14, 16]. However, the emulsion electrospinning would still cause damage or degradation of unstable macromolecules, like nucleic acids, mainly because of the shearing force and tension between the two phases of the emulsion, compared to coaxial electrospinning. Therefore, further modifications, like condensation of the carrier gene in gene therapy might be useful for more protection. Furthermore, during the emulsification or ultrasonication procedures in emulsion electrospinning, the contact of core materials with the solvent is increased, which may cause probable damage to the drug contents. Although extremely hydrophobic polymers can be used in emulsion electrospinning, the affinity or compatibility between drug and polymer might also influence the distribution of drugs within the fibers. It is cited that the copolymerization of hydrophobic polymers with hydrophilic polymers, such as PEG, ε-caprolactone (6-hexanolactone) (PCL), and poly(3-hydroxybutyric acid-co-3-hydroxyvaleric acid) (PHBV), affects drug distribution [3, 13, 14].

2.2.2.4 Surface modification electrospinning

The surface modification electrospinning is another promising method for introducing biofunctionality into nanofibers. In the surface modification electrospinning, a specific conductive surface can be chemically altered and changed aiming at modifying the external properties of a coated device, such as the tissue, which encircles the implanted material [3]. In this strategy, the release of the therapeutics is weakened and the functionality of the surface, where the immobilized biomolecules are located, preserved [13, 17]. Thus, this method is applied to avoid fast initial burst release and to slow down the rate of immobilization of the biological molecules on a particular surface. Therefore, the surface modification electrospinning is more applicable for gene or growth-factor delivery where slow and prolonged release of the therapeutic agent is required [13, 17]. Moreover, having a good electrospinning system and a well-standardized method, it is possible to coat 3D surfaces with nanoparticles or homogeneous surfaces [3, 16]. In cases where the drug cannot be immobilized, either because the drug is required to be endocytosed or interact with the nucleus of the cell, its release rate could be accurately controlled by introducing responsive materials to local external cues. This can happen by introducing hydrophobic functional groups onto the nanofibers surface [13].

2.3 Electrospinning parameters

The fabrication of nanofibers *via* electrospinning is affected by many different, but interlinked parameters as shown in **Table 1** [1]. These parameters modulate both the electrospinning process and the morphology of nanofibers [1, 4]. The electrospinning parameters can be classified as solution properties, process parameters, and environmental conditions [1, 4, 8]. The solution properties include the polymer concentration, molecular weight and viscosity, the solution conductivity and relative volatility, volatility of the solvent, surface tension, and dielectric constant. The process parameters refer to the flow rate, the applied voltage, the needle diameter, the distance between the tip of the needle and collector, and the geometry of the collector. The environmental conditions include the relative humidity and

Electrospinning parameters	Symbol	Effects on nanofiber morphology, diameter and its structure	
		Effects the morphology and structure	Effects on the diameter
Solution properties			
Concentration	C↑	Increasing concentration leads to increase in fiber diameter.	Nanofiber diameter↑
Viscosity	η↑	Increasing viscosity leads to thicker nanofibers without beads, but too high viscosity causes generation of beads.	Nanofiber diameter↑
Solution conductivity	σ↑	Increasing conductivity leads to thinner nanofibers.	Nanofiber diameter↓
Surface tension	γ	No conclusive correlation has been established between the surface tension and the nanofiber morphology.	-
Molecular weight of polymer	Mr↑	Increasing polymer molecular weight leads to formation of a nanofiber with fewer beads.	-
Volatility of solvent	$\alpha_{solvent}$	Higher volatility requires higher flow rate and leads to formation of a nanofiber with fewer beads.	-
Solution relative volatility	α	Porous microstructure appears because of higher volatility.	-
Dielectric constant	E_r	Sufficient dielectric constant of the solvent is needed for successful electrospinning.	-
Process parameters			
Flow rate	Q↑	Higher flow rate results in thicker nanofibers. Too high flow rate causes the generation of beads.	Nanofiber diameter↑
Applied voltage	V↑	Higher applied voltage leads to thinner nanofibers.	Nanofiber diameter↓
Needle diameter	D_{needle}		
Needle tip to collector distance	D↑	Minimum distance required to obtain dry nanofibers. Beaded morphology occurs when the distance is too short and the electric field is too strong.	Nanofiber diameter↓
Geometry of collector	-	Metal collectors with conductive frame or rotating drum are preferred.	-
Environmental conditions			
Relative humidity	φ	Porous microstructure appears due to evaporation-cooling effects. Lower humidity enables higher flow rate and the generation of beads is reduced.	-
Temperature	T↑	Higher temperature leads to thinner nanofibers.	Nanofiber diameter↓

Table 1.
Parameters that affect the electrospinning technique.

temperature [1, 4, 7, 8, 18]. The solution properties and the process parameters have predominant influence on the formation and morphology of the produced nanofibers, while the environmental conditions do not have a significant effect [1]. Moreover, all these factors are responsible for a flawless electrospinning process, which leads in the formation of the desirable electrospun nanofibers with the requisite characteristics [8]. Consequently, the careful monitoring of these factors can ensure the formation of smooth, highly porous nanofibers without beads [4, 8].

2.3.1 Effects of polymer concentration

The electrospinning method relies on the creation of electric charges in the polymer solution, which generate a charged jet [4]. When the polymer concentration is low, then the entangled polymer chains break into fragments before reaching the collector, due to the applied electric field and surface tension [4, 9, 12]. The entanglement of the polymer is necessary for fiber formation, but in the state of low polymer concentration. In this state, the phenomenon of electrospraying will take place and particles, instead of fibers, are formed [8, 19]. It has been reported that the boundary concentration between electrospray and electrospinning is solvent-dependent [8, 20]. Moreover, these polymer fragments cause the formation of nanofibers with beads [4]. In turn, if the polymer concentration increases, the chain entanglement among polymer chains will increase because of the increase in solution viscosity. As a result, these chain entanglements overcome the surface tension and uniform electrospun nanofibers devoid of beads are formed [4, 21]. If the concentration increases beyond a critical value, then the flow of the jet will be blocked as the droplet will dry out at the tip of the metallic needle, and the polymer jet would not be initiated. In this case, the clog should be removed to let the electrospinning process continue [4, 12, 18] and obtain beadles nanofibers with increased diameter [1, 4, 18, 22, 23].

2.3.2 Effects of polymer viscosity

With respect to the electrospinning method, the polymer viscosity is included in the solution properties. It has been reported that a change in polymer viscosity can affect the morphologies of the beads in nanofibers [4, 24]. If the viscosity of the polymer solution is low, the shape of the produced nanofibers will be round droplet like, but if the viscosity of the polymer solution is sufficient, then stretched droplet or eclipsed shapes fibers will be formed [4, 22–25]. Moreover, an increase in polymer concentration causes increase in polymer viscosity, and as a result, an increase beyond a critical value will block the flow of the jet and the droplet will dry out at the tip of the metallic needle. In conclusion, the determination of the critical value of viscosity is essential, as an increase in the polymer viscosity leads to thicker and bead-free nanofibers with increased diameter [4, 21]. Conversely, if the increase of viscosity is too high, beads will be generated in the nanofibers [1, 7].

2.3.3 Effects of solution conductivity

The solution conductivity is another solution parameter, which affects the electrospinning process and as a result the formation of nanofibers and their diameter distribution [7, 8]. The solution conductivity has a significant role on the formation of the Taylor cone and in controlling the diameter of the nanofibers [4, 8]. Poor conductivity solutions are not capable of producing electrospinning results, as the surface of the droplet will have no charge to form the Taylor cone. Conversely, an increase in the solution conductivity will lead to the Taylor cone formation because of the increase of the charge on the surface of the droplet. This will also lead to the reduction of the fiber diameter [4, 8, 26]. It has been reported that if the solution conductivity increases beyond a critical value, the formation of the Taylor cone will be prevented. This can be attributed to the Coulombic forces between the charges on the surface of the fluid and the force due to the external electric field [4]. It has been well documented that a highly conductive polymer solution is unstable and leads to a wide diameter distribution when a strong electric field is applied [4, 7, 27]. However, the polymer solution conductivity could be adjusted by the addition of a suitable salt [4, 7]. The addition of the salt affects the electrospinning process by increasing the number of the ions in the polymer solution resulting in the increase of surface charge density of the fluid and the electrostatic force produced by the applied electric field [4, 7, 22, 28, 29]. Moreover, the addition of the salt increases the polymer solution conductivity resulting in the reduction in the tangential electric field along the surface of the fluid [4]. Concluding, the increase of the solution conductivity leads to ultrafine nanofibers with reduced diameter [1, 7, 8, 26].

2.3.4 Effects of surface tension

The surface tension is included in the solution parameters, which affect the electrospinning process and the nanofiber morphology, but there is no conclusive correlation [1]. Nevertheless, it has been reported that there is a delicate balance between the surface tension and the electric field (conductivity, concentration, and viscosity), which affects the ultimate morphology of the nanofibers [4, 7]. Particularly, the surface tension and the applied electric field cause the disentangling and breaking of the perplexed polymer chains into fragments before reaching the collector, which cause the formation of beads in the nanofibers [4, 9, 12]. Another case refers that the surface tension influences the surface of the polymeric nanofibers, and in the case of poor conductivity of polymers, charges accumulate onto the surface and as a result, beaded formation is prompted [7].

2.3.5 Effects of molecular weight of polymer

The molecular weight of the polymer is included in the solution properties, and it is a parameter that affects the viscosity of the solution. Ordinarily, an increase in molecular weight, until a critical value, leads to increase in solution viscosity and the formation of nanofibers with fewer beads [1, 7]. In general, polymers with high molecular weight are preferred as they cause extensive chain entanglement, which facilitates the nanofiber formation during the spinning process. On the contrary, polymer solutions with lower molecular weight may lead to the formation of beads or break up into droplets [30]. Overall, the molecular weight is one of the most important parameters, which affect the outcome nanofibers and as a result the electrospinning process.

2.3.6 Effects of solvent volatility

The solvent volatility is another parameter of solution parameters, which affects the electrospinning process and as a result the formation of smooth and beadles electrospun nanofibers. The solvents that are preferred in the electrospinning process should be polymers that are entirely soluble, and they should have moderate (appropriate) boiling point, which is related with the volatility of the solvent [4, 8]. Common volatile solvents, with high evaporation rates, which ensure the facile evaporation of the solvent from the tip of the needle to collector, are used in the electrospinning process [4]. The rate of solvent evaporation from the polymer solution jet leads to phase separation and creation of secondary structures on fibers [4, 7, 31]. It has been reported that highly volatile solvents absorb the heat from the jet, thus lowering the temperature of the liquid jet; this temperature drop decreases the thermodynamic stability of the nonsolvent phase. These phenomena result in high evaporation rates, which cause the drying of the jet at the tip of the needle, block the needle tip, and hence hinder the electrospinning process or else the early solidification of the polymer jet. Overall, highly volatile solvents are avoided in the electrospinning process because fiber formation will not be completed [4, 7, 8]. Similarly, solvents with low volatility should not be used, because they have high boiling points, which prevent the drying during the nanofiber jet formation or else the solidification process could be retarded because the solvent evaporation is low [4, 8]. Conclusively, the type of the solvent and especially their volatility profile, and the rate of evaporation are very important parameters for the formation of nanofibers. It is cited that higher volatility demands and higher flow rates result in the formation of electrospun nanofibers with fewer beads [1, 4, 7].

2.3.7 Effects of solution volatility

Relative volatility is a measure of the differences in volatility between two components and is used in the design of separation or absorption processes. The solution relative volatility is a solution parameter that has similar effect with the volatility of the solvent. Solutions that are prepared from solvents of very low volatility may deliver wet and cross-linked nanofibers or even no nanofibers, at all [4, 8, 30]. Conversely, the usage of highly volatile solvents for the solution preparation may result in intermittent spinning because of the solidification of the polymer jet at the tip of the needle [4, 7, 30]. It has been reported that an increase in relative volatility of polymer solution causes the appearance of porous microstructure due to higher volatility, and this affects fiber's porosity and morphology [7].

2.3.8 Effects of dielectric constant

The dielectric constant, sometimes called relative permittivity or specific inductive capacity, is the ratio of the permittivity of a substance to the permittivity of free space. It is an expression of the extent to which a material concentrates electric flux. The dielectric constant of the solvent(s), used in the successful electrospinning and the formation of electrospun nanofibers, has to be sufficient, but not high [1]. It has been reported that an increase in the dielectric constant of the solution leads to an increase of the number of jets. On the contrary, a reduction in the value of the dielectric constant to a single digit leads to the formation of a single jet. Furthermore, the value of the solution dielectric constant may influence the stability of the jet, as bending instability may be reduced with a lower charge density resulting in a longer and stable jet [30]. Overall, the solution dielectric constant has to be sufficient for the successful electrospinning and the formation of electrospun nanofibers [1].

2.3.9 Effects of flow rate

The flow rate is an important parameter belonging to the process parameters, which influences the diameter of the electrospun fibers and subsequently the charge density and the morphology of the nanofibers [4, 7]. It is reported that there is a critical point depending on the polymeric solution, in which the critical flow rate leads to the formation of uniform electrospun nanofibers [4, 8]. In the case of increasing the flow rate, beyond the critical value, nanofibers with larger diameter and pore size are produced and the formation of beaded structures is enhanced [4, 7, 8, 18, 31]. This bead formation is caused due to the incomplete drying of the polymeric jet. When the delivery rate of the polymeric jet to the needle tip exceeds the rate at which the polymeric solution is removed from the tip by the electric force in the metallic collector, a mass balance shift results, which leads to a sustained but unstable jet and bead formation [4, 7, 32]. In the case of decreasing the flow rate, beyond the critical value, smooth, fine, and thinner nanofibers are formed [1, 18]. It is cited that increases and decreases in the flow rate affect the nanofiber formation, and as a result, a minimum flow rate of the polymeric solution is preferred in order to replace the solution that is lost with a new one, during jet formation, as the solution will have enough time for polarization, stretching, and drying [4, 31]. Overall, lowering the flow rate causes the formation of thinner nanofibers instead of too high flow rates in which the nanofiber diameter increases and the continuity of the fiber interrupts bead formation [1, 7].

2.3.10 Effects of applied voltage

The applied voltage is an important process parameter, which affects the strength of the electric field and therefore influences the diameter and the nanofiber morphology [7, 8]. Moreover, an increase in the applied voltage causes a change in the shape of the Taylor cone, and as a result, a critical voltage, which depends on the polymeric solution, is needed for the formation of ultrafine nanofibers given a certain distance between the needle tip and collector [4, 5, 7, 8]. An increase in the applied voltage leads to the formation of thinner nanofibers because of the stretching of the polymer solution in correlation with the charge repulsion within the polymer jet [1, 4, 7, 18, 33]. A higher applied voltage, above the critical value, may lead to an irregular increase of the diameter and the formation of beaded, nonuniformity nanofibers [4, 7, 8, 22]. This situation is attributed to the decrease in the size of the Taylor cone and increase in the jet velocity, keeping the same flow rate [4, 22, 34]. However, there are studies that have shown that the increase in the applied voltage leads to increase in the diameter of the nanofibers [4, 18, 21]. This phenomenon may be explained as the increase of the voltage leads to the decrease of the volume of the drop at the tip of the needle causing the receding of the Taylor cone resulting in increase in the jet length and fiber diameter because of the increase in the amount of the ejected fluid and the flow rate of polymer solution [4, 18, 21]. In conclusion, in general, the increase of the applied voltage, until a critical value, causes the formation of thinner nanofibers, but this depends on the type of the polymeric solvent [1, 4, 7, 8, 18]. It is worth mentioning that the problem of bead formation was not solved by varying the applied voltage [18].

2.3.11 Effects of needle tip to collector distance

The distance between the metallic needle tip and the collector could be easily affecting the morphology of nanofibers because it is dependent on the deposition time, evaporation rate, and the whipping or instability interval [4, 7, 8, 35]. Therefore, a critical distance is needed to be fixed for the preparation of dry, smooth, and uniform electrospun nanofibers [1, 4]. A decrease in the distance between the tip and the collector leads to the enlargement of diameter of nanofibers and the generation of beads, while an increase in this distance leads to the formation of nanofibers with decreased diameter [1, 7, 8, 21, 35]. However, there are cases that the morphology of nanofibers is not affected by the distance between the metallic needle and the collector [4, 32]. Increasing the distance between the needle tip and the collector, the nanofiber diameter decreases and there is a minimum distance required to obtain dry, smooth, and uniform electrospun nanofibers, but when the distance is too short or too large, beads are formed [1, 7].

2.3.12 Effects of relative humidity

The relative humidity is a factor belonging to the environmental conditions of the electrospinning, which affects the diameter and the morphology of the electrospun nanofibers [4, 8, 36, 37]. The relative humidity is crucial for the production of ultrafine nanofibers with acceptable morphology, because it affects the formation of pores on the fiber surface *via* solvent evaporation or else controlling the solidification process of the charged jet [4, 7, 8]. The appropriate amount of the relative humidity depends on the chemical nature of the used polymer. A high relative humidity suppresses the evaporation rate as long as the surface area of the jet increases and the charge per unit area on the surface of the jet decreases resulting in the capillary instability and the beaded structure formation [1, 7, 8]. It has been cited that humidity controls the evaporation rate of the fluid jet when the water is used as a solvent component [7]. Overall, lower relative humidity enables higher flow rate, and as a result, the formation of beads is reduced, while higher relative humidity leads to the appearance of porous microstructures due to evaporation effects and/or phase separation [1, 7, 8].

2.3.13 Effects of temperature

The temperature is another factor belonging to the environmental conditions of the electrospinning, which is crucial for the production of ultrafine nanofibers with acceptable morphology, because it affects the diameter of the fibers [4, 8, 36, 37]. Moreover, temperature causes changes in the average diameter of the nanofibers resulting in modification of the electrospun nanofibers size by causing two opposing effects; first, it increases the evaporation rate of the solvent and secondly, it decreases the viscosity of the polymer solution. These effects have the behavior of two opposite mechanisms, but both of them lead to a mean fiber diameter decrease [4, 8]. In general, an increase in the temperature leads to thinner nanofibers formation [1, 4, 8].

3. Electrospinning in *per* oral drug delivery

With the appearance of nanotechnology, researchers have become more attracted in studying the characteristic properties of nanoscale materials. Electrospinning, a method of electrostatic fiber fabrication, has established more attention in recent years due to its usefulness and potential for applications in diverse fields, like tissue engineering, biosensors, filtration, wound dressings, drug delivery, and enzyme immobilization. The nanoscale fibers are generated by the application of strong electric fields on polymer solution and mimic better the extracellular matrix components as compared to the conventional techniques, offering various advantages, like high surface area to volume ratio, tunable porosity, and the ability to manipulate nanofiber composition in order to get desired properties and function [38]. The use of electrospun nanofibers, as formulation systems for oral drug delivery, has been studied extensively over the past decades in fast/immediate release systems and more recently in modified release systems.

Numerous researchers have been studying orodispersible or fast-dissolving drug delivery formulations produced from nanofiber-loaded systems that rapidly disintegrate in the oral cavity due to nanofibers' large surface area, which causes immediate disintegration in water solutions and fast drug release [18, 39–46]. Applications of the electrospinning technique on modified *per* oral drug delivery are summarized in **Table 2**.

3.1 Electrospinning in controlled *per* oral drug delivery

Oral controlled drug release systems are characteristic in formulation, and researchers have developed electrospun nanofibers for usage in treatment and management of disorders that need special drug release patterns. Scientists have developed amyloid-like bovine serum albumin with ampicillin sodium salt nanofibers by electrospinning, and the *in vitro* results showed controlled release behavior [47]. Electrospun fiber mats were also investigated as drug delivery systems using tetracycline hydrochloride as a model drug. The nanofibers were made either from poly(lactic acid), poly(ethylene-co-vinyl acetate), or from a 50:50 blend of the two. The release of the tetracycline hydrochloride from these new drug delivery systems followed controlled release behavior [48]. Moreover, polyvinyl alcohol nanofibers loaded with curcumin or its β-cyclodextrin inclusion complexes were prepared using an electrospinning process. *In vitro* dissolution tests showed that the drug release profiles of polyvinyl alcohol/curcumin and polyvinyl alcohol/complex fibers were different, with release from the latter occurring more rapidly [49]. In addition, electrospun gelatin nanofibers were prepared by sequential crosslinking

drug release behavior	delivery system	API	excipient (s)	electrospinning technique	ref
controlled	mats	ampicillin sodium salt	amyloid-like bovine serum albumin	blending	47
	mats	tetracycline hydrochloride	poly(lactic acid), poly(ethylene-co-vinyl acetate)	blending	48
	mats	curcumin	polyvinyl alcohol, β-cyclodextrin	blending	49
	multilayered gelatin mesh	piperine	gelatin (type A), acetic acid	multiple blending with sequential crosslinking using glutaraldehyde	50
	mats in hard gelatin capsules	melatonin	cellulose acetate, polyvinylpyrrolidone and hydroxypropylmethylcellulose	blending	51
	mats in 3-layered tablets	melatonin	cellulose acetate polyvinylpyrrolidone	blending	52
delayed	mats	5-fluorouracil	Core: poly(vinylpyrrolidone), ethyl cellulose, methacrylic acid copolymer S100 or drug alone Shell: methacrylic acid copolymer S100	coaxial	53
	gelatin nanofibers	piperine	gelatin (type A), acetic acid	sequential crosslinking using glutaraldehyde	54
	nano-fiber packed capsules	uranine and nifedipine	methacrylic acid copolymer S100	blending	55
	mats in tablets	acetaminophen	methacrylic acid copolymer S100	blending	56
colon targeted	mats	diclofenac sodium	methacrylic acid copolymer L100-55	blending	57
	mats	ferulic acid	Core: shellac Shell: N,N-dimethylformamide	coaxial	58
	mats	indomethacin	methacrylic acid copolymer RS100 and S100	blending	59
	mats	indomethacin	methacrylic acid copolymer RS100 and S100	blending	60
	mats	celecoxib	pectin, methacrylic acid copolymer RS30D, polycaprolactone	blending	61
	milled mats	budesonide	methacrylic acid copolymer S100	blending	62
	nanofilm	bovine serum albumin	Core: chitosan Shell: alginate	coaxial	63
	mats	doxorubicin	polydopamine, poly-ε-caprolactone	blending	64
biphasic	mats	ketoprofen	Core: ethyl cellulose Shell: polyvinylpyrrolidone	coaxial	65
	mats	ketoprofen	Core: zein Shell: polyvinylpyrrolidone	coaxial	66
	tri-layered mesh	ketoprofen	Core: zein Shell: polyvinylpyrrolidone and graphene oxide	sequential coaxial	67
	gelatin coated	ciprofloxacin	Mg-Ca alloy	blending with crosslinking using glutaraldehyde	68
	mats	resveratrol	polycaprolactone	blending	69
	mats	ampicillin	Core/Shell: polycaprolactone	coaxial	70
	mats	piroxicam	hydroxypropylmethylcellulose	blending	71
dual	mats	aceclofenac/ pantoprazole	zein/methacrylic acid copolymer S100	blending	72

Table 2.
An overview of the electrospinning technique applications in modified ***per*** *oral drug delivery.*

using piperine as a hydrophobic model drug by sandwiching the drug-loaded gelatin nanofiber mesh with another gelatin nanofiber matrix without drug (acting as diffusion barrier). The results indicated controlled and sustainable release of the drug for prolonged time [50]. Researchers have also prepared melatonin-loaded nanofibrous systems based on cellulose acetate, polyvinylpyrrolidone, and hydroxypropylmethylcellulose. The electrospun nanofiber mats that were inserted in hard gelatin capsules exhibited variable release profiles in the gastric-like fluids, ranging from 30 to 120 min, while the electrospun nanofiber mats that were inserted in DRcaps™ capsules released melatonin at a slower pace [51]. In another study, nanofibers of cellulose acetate and polyvinylpyrrolidone loaded with melatonin were prepared and compressed at various pressures into monolayered tablets. The nanofiber mats were then incorporated into three-layered tablets, containing in the upper and lower layer combinations of lactose monohydrate and hydroxypropylmethylcellulose, as modifying accessories, and their *in vitro* dissolution profiles have showed promising results in modified *per* oral drug delivery [52].

3.2 Electrospinning in delayed *per* oral drug delivery

Besides controlled drug release, researchers have investigated electrospun nanofibers as oral delivery systems for delayed release systems. In a study, both fast dissolving and sustained release drug delivery systems comprising mebeverine hydrochloride embedded in either povidone K60 or Eudragit RL 100–55 nanofibers have been prepared by electrospinning. The *in vitro* dissolution tests of the povidone K60 fiber mats revealed dissolution within 10 s, while the Eudragit fibers revealed pH-dependent drug release profiles, with only very limited release at pH 2.0, but sustained release over approximately 8 h at pH 6.8. As a result, it can be stated that the Eudragit nanofibers have the potential to be developed as oral drug delivery systems for localized drug release in the intestinal tract, whereas the povidone materials may find application as buccal delivery systems or suppositories [53]. Various researchers have synthesized gelatin nanofibers by electrospinning, using piperine as a hydrophobic model drug. The electrospun gelatin nanofibers were cross-linked by exposing to saturated glutaraldehyde vapor, to improve their water-resistive properties. The results illustrated good compatibility of the hydrophobic drug in gelatin nanofibers with promising controlled drug release patterns by varying cross-linking time and the pH of the release medium [54]. In another scientific report, a solvent-based electrospinning method was used to prepare nanofiber-based capsules including drugs (uranine was used as a water-soluble drug and nifedipine as a water-insoluble drug) for controlled release delivery systems using methacrylic acid copolymer as a polymer. The *in vitro* release of uranine or nifedipine from the nanofiber-packed capsules and milled powder of nanofiber-packed capsules showed controlled release of uranine or nifedipine, as compared to capsules of a physical mixture of methacrylic acid copolymer and each drug. The *in vivo* pharmacokinetic evaluation in rats, after intraduodenal administration of nanofiber-packed capsules or milled powder of nanofiber-packed capsules including uranine and/or nifedipine, clearly demonstrated that the application of the nanofibrotic technique, as a drug delivery system, offers drastic changes in pharmacokinetic profiles for both water-soluble and water-insoluble drugs [55]. Furthermore, nanofibers made from methacrylic acid copolymer S, containing acetaminophen, were prepared using a solvent-based electrospinning method. The *in vitro* dissolution rate profiles of acetaminophen showed that the tablets based on methacrylic acid copolymer S nanofibers did not disintegrate in the intestine in the lower pH region and could regulate the drug release in a pH-dependent manner [56].

3.3 Electrospinning in colon-targeted *per* oral drug delivery

In addition to the previously described drug delivery systems, many scientists have demonstrated that the electrospinning method could be regarded as a modern approach for the preparation of colon drug delivery systems leading to marketable products. Eudragit L 100-55 nanofibers loaded with diclofenac sodium were successfully prepared using an electrospinning process. *In vitro* dissolution tests verified that all the drug-loaded Eudragit L 100-55 nanofibers had pH-dependent drug release profiles, with limited release at pH 1.0, but a sustained and complete release at pH 6.8, indicating the potential of oral colon-targeted drug delivery systems development [57]. Researchers prepared medicated shellac nanofibers providing colon-specific sustained release of ferulic acid using coaxial electrospinning. The *in vitro* dissolution tests demonstrated that there was minimal ferulic acid release at pH 2.0, and sustained release in a neutral dissolution medium [58]. Another group of researchers have prepared electrospun nanofibers of indomethacin aimed for colon delivery using Eudragit S and Eudragit RS as polymers. It was shown that the ratio of drug:polymer and polymer:polymer were pivotal factors to control the drug release from nanofibers. A formulation containing Eudragit S:Eudragit RS (60:40) and drug:polymer ratio of 3:5 exhibited the most appropriate drug release, as a colon delivery system with a minor release at pH 1.2, 6.4, and 6.8 and a major release at pH 7.4 [59]. Electrospun nanofibers were also successfully prepared using indomethacin as a drug and Eudragit RS100 and S100 as polymers for colonic drug delivery [60]. Moreover, celecoxib-loaded electrospun nanofibers were developed using a combination of time-dependent polymers with pectin to achieve colon-specific drug delivery systems. The drug release was limited in the acidic media; while, in the simulated colonic media, it was higher from formulations containing the excipient pectin [61]. Likewise, electrospun fibers loaded with budesonide were prepared with the aim of controlling its release in the gastrointestinal tract using Eudragit RS 100, a polymer soluble at pH > 7, commonly used for enteric release of drugs. The dissolution rate measurements using a pH-change method showed low drug dissolution at pH 1.0 and sustained release at pH 7.2, representing an effective method for drug targeting to terminal ileum and colon with the aim of improving the local efficacy of budesonide for the treatment of some inflammatory bowel diseases [62]. Researchers have developed a novel core-shell-structured nanofilm for colon delivery by coaxial electrospinning using bovine serum albumin as protein model. First, the protein-loaded chitosan nanoparticle was prepared by ionic gelation, and then, the coaxial nanofilm was fabricated using alginate as shell layer and the protein-loaded chitosan nanoparticle as core layer. The protein release in different simulated digestive fluids revealed that the electrospun nanofilm is a promising colon-specific delivery system for bioactive proteins [63]. Another group of scientists reported in their work that the pH-responsive drug delivery systems could mediate drug releasing rate by changing the pH values at specific times as *per* the pathophysiological need of the disease. Their study demonstrated that a mussel-inspired protein polydopamine coating can tune the loading and releasing rate of charged molecules from electrospun poly(ε-caprolactone) nanofibers in solutions with different pH values. The *in vitro* release profiles showed that the positively charged molecules led to a significantly faster release in acidic than in neutral and basic media, while the results of specialized assays showed that the media containing doxorubicin released in solutions at low pH values could kill a significantly higher number of cells than those released in solutions at higher pH values. The pH-responsive drug delivery systems based on polydopamine-coated poly(e-caprolactone) nanofibers could have potential application in the oral delivery of anticancer drugs for treating gastric cancer and in vaginal delivery of antiviral or anti-inflammatory drugs, which could raise their efficacy, deliver them to the specific site, and minimize their toxicity [64].

3.4 Electrospinning in biphasic and dual *per* oral drug delivery

More to the point of modified drug delivery systems, researchers have designed and fabricated nanostructures using electrospinning for providing biphasic drug release profiles. A research work investigated the biphasic release profile of ketoprofen of core/sheath nanofibers prepared using as polymers polyvinylpyrrolidone for the sheath and ethyl cellulose for the core matrix by coaxial electrospinning. The *in vitro* dissolution tests showed that the nanofibers produced could provide a biphasic drug release profile consisting of an immediate and a sustained release [65]. In another work, core-sheath nanofibers were also prepared using ketoprofen as a model drug, and polyvinylpyrrolidone and zein as the sheath polymer and core matrix excipient, respectively, by coaxial electrospinning. The *in vitro* dissolution tests showed that the nanofibers could provide an immediate release of 42.3% of the drug followed by a sustained release over 10 h of the remaining drug [66]. Other researchers have used simple sequential electrospinning to create a triple layered nanofiber mesh with biphasic drug release behavior. The mesh was composed of zein and polyvinylpyrrolidone as the top/bottom and middle layers, respectively. Ketoprofen was used as a model drug, and polyvinylpyrrolidone was blended with graphene oxide to improve the drug release functionality of the nanofiber as well as its mechanical properties. The *in vitro* release tests demonstrated time-regulated biphasic drug release [67]. In another study, gelatin-ciprofloxacin nanofibers containing various amounts of ciprofloxacin were fabricated on the surface of Mg-Ca alloy *via* an electrospinning process. Prolonged drug release was attained from gelatin-ciprofloxacin nanofibers coating along with initial rapid drug release of around 20–22% during 12 h, followed by a slow release stage that can effectively control the infection [68]. Moreover, resveratrol (a promising natural substance for periodontal disease treatment due to its anti-inflammatory and antioxidative effects) was successfully incorporated into polycaprolactone-nanofibers and enabled a biphasic-release kinetic pattern [69]. In a recent study, it was demonstrated that the production of core-shell fibers via modified coaxial electrospinning achieved controlled release of ampicillin-loaded polycaprolactone nanofibers covered by a polycaprolactone shield. The *in vitro* release studies showed that the drug release kinetics of core-shell products is closer to zero-order kinetics, while the drug release kinetics of single electrospinning of the core resulted with burst release [70]. Scientists have also used piroxicam as a low-dose, poorly soluble drug and hydroxypropyl methylcellulose as an amorphous-state stabilizing carrier polymer in nanofibers to produce biphasic-release drug delivery systems [71].

Dual drug delivery systems have also been successfully developed by researchers. In a recent study, aceclofenac/pantoprazole-loaded zein/Eudragit S 100 nanofibers were developed using a single nozzle electrospinning process. The *in vitro* release studies ensured the efficiency of the nanofibers in sustaining the release of both drugs up to 8 h, while the *in vivo* experiments confirmed that the co-administration of pantoprazole and aceclofenac reduced the gastrointestinal toxicity induced by nonsteroidal anti-inflammatory drugs [72].

4. Conclusions

The fabrication of electrospun ultrafine fiber meshes from biodegradable and biocompatible polymers has opened new horizons in the biomedical field. Electrospinning, being a simple, highly versatile, and robust technique for the production of fibers with diameters from submicron down to nanometer scale, could provide a useful method for the development of novel drug carriers capable of affecting the drugs' modified release. By careful selection of polymers, it is now possible to deliver drugs, with diverse stereoelectronic and physicochemical properties, in a required manner using electrospun nanofibers. *Mutatis mutandis*, in order to make further progress in the drug delivery field, it is necessary to identify ways that will allow fabrication of nanofibers with the desired morphological and mechanical properties in a reproducible manner. Thus, organic solvent mixtures, drug content, and electrospinning parameters, which will influence nanofiber properties, such as morphology, applicability, and quality, are currently under intense investigation.

Conflict of interest

The authors declare no conflict of interest.

Author details

Marilena Vlachou*, Angeliki Siamidi and Sotiria Kyriakou
Department of Pharmacy, Division of Pharmaceutical Technology, School of Health Sciences, National and Kapodistrian University of Athens, Athens, Greece

*Address all correspondence to: vlachou@pharm.uoa.gr

References

[1] Sunil CU, Shridhar NB, Jagadeesh SS, Ravikumar C. Nanofibers in drug delivery: An overview. World Journal of Pharmaceutical Research. 2015;**4**(8):2576-2594. ISSN: 2277-7105

[2] Sujitha R, Moin A, Gowda DV, Jigyasa V, Santhosh TR, Osmani RAM. Nanofibers: The new-fangled loom in drug delivery and therapeutics. Indo American Journal of Pharmaceutical Research. 2016;**6**(03):4690-4697. ISSN: 2231-6876

[3] Manuel CBJ, Jesus VGL, Aracely SM. Electrospinning for drug delivery systems: Drug incorporation techniques. In: Haider S, Haider A, editors. Electrospinning - Material, Techniques, and Biomedical Applications. London: IntechOpen; 2016. pp. 141-155. DOI: 10.5772/65939

[4] Haider A, Haider S, Kang IK. A comprehensive review summarizing the effect of electrospinning parameters and potential applications of nanofibers in biomedical and biotechnology. Arabian Journal of Chemistry. 2018;**11**:1165-1188. DOI: 10.1016/j.arabjc.2015.11.015

[5] Laudenslager MJ, Sigmund WM. Electrospinning. In: Bhushan B, editor. Encyclopedia of Nanotechnology. Springer: Dordrecht; 2012. pp. 769-775. DOI: 10.1007/978-90-481-9751-4_357

[6] Wang B, Wang Y, Yin T, Yu Q. Applications of electrospinning technique in drug delivery. Chemical Engineering Communications. 2010;**197**(10):1315-1338. DOI: 10.1080/00986441003625997

[7] Weng L, Xie J. Smart electrospun nanofibers for controlled drug release: Recent advances and new perspectives. Current Pharmaceutical Design. 2015;**21**(15):1944-1959. DOI: 10.2174/1381612821666150302151959

[8] Akhgari A, Shakib Z, Sanati A. A review on electrospun nanofibers for oral drug delivery. Nanomedicine Journal. 2017;**4**(4):197-207. DOI: 10.22038/nmj.2017.04.001

[9] Pillay V, Dott C, Choonara YE, Tyagi C, Tomar L, Kumar P, et al. A review of the effect of processing variables on the fabrication of electrospun nanofibers for drug delivery applications. Journal of Nanomaterials. 2013;**2013**:1-22. DOI: 10.1155/2013/789289

[10] Braghirolli DI, Steffens D, Pranke P. Electrospinning for regenerative medicine: A review of the main topics. Drug Discovery Today. 2014;**19**(6):743-753. DOI: 10.1016/j.drudis.2014.03.024

[11] Bae HS, Haider A, Selim KMK, Kang DY, Kim EJ, Kang IK. Fabrication of highly porous PMMA electrospun fibers and their application in the removal of phenol and iodine. Journal of Polymer Research. 2013;**20**(7):1-7. DOI: 10.1007/s10965-013-0158-9

[12] Haider S, Al-Zeghayer Y, Ahmed Ali F, Haider A, Mahmood A, Al-Masry W, et al. Highly aligned narrow diameter chitosan electrospun nanofibers. Journal of Polymer Research. 2013;**20**(4):1-11. DOI: 10.1007/s10965-013-0105-9

[13] Zamani M, Prabhakaran PM, Ramakrishna S. Advances in drug delivery via electrospun and electrosprayed nanomaterials. International Journal of Nanomedicine. 2013;**8**(1):2997-3017. DOI: 10.2147/IJN.S43575

[14] Imani R, Yousefzadeh M, Nour S. Functional nanofiber for drug delivery applications. In: Barhoum A, Bechelany M, Makhlouf A, editors. Handbook of Nanofibers. Cham: Springer; 2018. pp. 1-55. DOI: 10.1007/978-3-319-42789-8_34-1

[15] Tipduangta P, Belton P, Fábián L, Wang LY, Tang H, Eddleston M, et al.

Electrospun polymer blend nanofibers for tunable drug delivery: The role of transformative phase separation on controlling the release rate. Molecular Pharmaceutics. 2016;**13**(1):25-39. DOI: 10.1021/acs.molpharmaceut.5b00359

[16] Ravi Kumar RMV. Handjournal of Polyester Drug Delivery Systems. 1st ed. Vol. 1. Boca Ratón, FL: CRC Press; 2016. pp. 1-738. ISBN: 9789814669658

[17] Volpato FZ, Almodovar J, Erickson K, Popat KC, Migliaresi C, Kipper MJ. Preservation of FGF-2 bioactivity using heparin-based nanoparticles, and their delivery from electrospun chitosan fibers. Acta Biomaterialia. 2012;**8**(4):1551-1559. DOI: 10.1016/j.actbio.2011.12.023

[18] Reda RI, Wen MM, El-Kamel AH. Ketoprofen-loaded Eudragit electrospun nanofibers for the treatment of oral mucositis. International Journal of Nanomedicine. 2017;**12**:2335-2351. DOI: 10.2147/IJN.S131253

[19] Ahmed FE, Lalia BS, Hashaikeh R. A review on electrospinning for membrane fabrication: Challenges and applications. Desalination. 2015;**356**: 15-30. DOI: 10.1016/j.desal.2014.09.033

[20] Costa LMM, Bretas RES, Gregorio R. Effect of solution concentration on the electrospray/electrospinning transition and on the crystalline phase of PVDF. Materials Sciences and Applications. 2010;**1**:247-252. DOI: 10.4236/msa.2010.14036

[21] Baumgarten PK. Electrostatic spinning of acrylic microfibers. Journal of Colloid and Interface Science. 1971;**36**(1):71-79. DOI: 10.1016/0021-9797(71)90241-4

[22] Zong X, Kim K, Fang D, Ran S, Hsiao BS, Chu B. Structure and process relationship of electrospun bioabsorbable nanofiber membranes. Polymer. 2002;**43**(16):4403-4412. DOI: 10.1016/S0032-3861(02)00275-6

[23] Fong H, Chun I, Reneker D. Beaded nanofibers formed during electrospinning. Polymer. 1999;**40**(16):4585-4592. DOI: 10.1016/S0032-3861(99)00068-3

[24] Shamim Z, Saeed B, Amir T, Abo Saied R, Rogheih D. The effect of flow rate on morphology and deposition area of electrospun nylon 6 nanofiber. Journal of Engineered Fibers and Fabrics. 2012;**7**(4):42-49. DOI: 10.1177/155892501200700414

[25] Doshi J, Reneker DH. Electrospinning process and applications of electrospun fibers. Journal of Electrostatics. 1995;**35**(2-3):151-160. DOI: 10.1016/0304-3886(95)00041-8

[26] Sun B, Long YZ, Zhang HD, Li MM, Duvail JL, Jiang XY, et al. Advances in three-dimensional nanofibrous macrostructures via electrospinning. Progress in Polymer Science. 2014;**39**(5):862-890. DOI: 10.1016/j.progpolymsci.2013.06.002

[27] Hayati I, Bailey AI, Tadros TF. Investigations into the mechanisms of electrohydrodynamic spraying of liquids: I. Effect of electric field and the environment on pendant drops and factors affecting the formation of stable jets and atomization. Journal of Colloid and Interface Science. 1987;**117**(1):205-221. DOI: 10.1016/0021-9797(87)90185-8

[28] Cai S, Xu H, Jiang Q, Yang Y. Novel 3D electrospun scaffolds with fibers oriented randomly and evenly in three dimensions to closely mimic the unique architectures of extracellular matrices in soft tissues: Fabrication and mechanism study. Langmuir. 2013;**29**(7):2311-2318. DOI: 10.1021/la304414j

[29] Choi JS, Lee SW, Jeong L, Bae SH, Min BC, Youk JH, et al. Effect of organosoluble salts on the nanofibrous structure of electrospun

poly(3-hydroxybutyrate-co-3-hydroxyvalerate). International Journal of Biological Macromolecules. 2004;**34**(4):249-256. DOI: 10.1016/j.ijbiomac.2004.06.001

[30] Teo WE. Introduction to Electrospinning Parameters and Fiber Control. 1st ed. Singapore: ElectrospinTech; 2015. pp. 25-29

[31] Megelski S, Stephens JS, Bruce Chase D, Rabolt JF. Micro- and nanostructured surface morphology on electrospun polymer fibers. Macromolecules. 2002;**35**(22): 8456-8466. DOI: 10.1021/ma020444a

[32] Zhang C, Yuan X, Wu L, Han Y, Sheng J. Study on morphology of electrospun poly(vinyl alcohol) mats. European Polymer Journal. 2005;**41**(3):423-432. DOI: 10.1016/j.eurpolymj.2004.10.027

[33] Sill TJ, von Recum HA. Electrospinning: Applications in drug delivery and tissue engineering. Biomaterials. 2008;**29**(13):1989-2006. DOI: 10.1016/j.biomaterials.2008.01.011

[34] Deitzel JM, Kleinmeyer J, Harris D, Beck Tan NC. The effect of processing variables on the morphology of electrospun nanofibers and textiles. Polymer. 2001;**42**(1):261-272. DOI: 10.1016/S0032-3861(00)00250-0

[35] Matabola KP, Moutloali RM. The influence of electrospinning parameters on the morphology and diameter of poly(vinyledene fluoride) nanofibers-effect of sodium chloride. Journal of Materials Science. 2013;**48**(16):5475. DOI: 10.1002/app.31396

[36] Huan S, Liu G, Han G, Cheng W, Fu Z, Wu Q, et al. Effect of experimental parameters on morphological, mechanical and hydrophobic properties of electrospun polystyrene fibers. Materials. 2015;**8**(5):2718. DOI: 10.3390/ma8052718

[37] Pelipenko J, Kristl J, Jankovic' B, Baumgartner S, Kocbek P. The impact of relative humidity during electrospinning on the morphology and mechanical properties of nanofibers. International Journal of Pharmaceutics. 2013;**456**(1):125-134. DOI: 10.1016/j.ijpharm.2013.07.078

[38] Bhardwaj N, Kundu SC. Electrospinning: A fascinating fiber fabrication technique. Biotechnology Advances. 2010;**28**(3):325-347. DOI: 10.1016/j.biotechadv.2010.01.004

[39] Akhgari A, Ghalambor Dezfuli A, Rezaei M, Kiarsi M, Abbaspour MR. The design and evaluation of a fast dissolving drug delivery system for loratadine using the electrospinning method. Jundishapur Journal of Natural Pharmaceutical Products. 2016;**11**(2):e33613. DOI: 10.17795/jjnpp-33613

[40] Illangakoon UE, Gill H, Shearman GC, Parhizkar M, Mahalingam S, Chatterton NP, et al. Fast dissolving paracetamol/caffeine nanofibers prepared by electrospinning. International Journal of Pharmaceutics. 2014;**477**(1-2):369-379. DOI: 10.1016/j.ijpharm.2014.10.036

[41] Li X, Kanjwal MA, Lin L, Chronakis IS. Electrospun polyvinylalcohol nanofibers as oral fast-dissolving delivery system of caffeine and riboflavin. Colloid Surface B. 2013;**103**:182-188. DOI: 10.1016/j.colsurfb.2012.10.016

[42] Nam S, Lee JJ, Lee SY, Jeong JY, Kang WS, Cho HJ. Angelica gigas Nakai extract-loaded fast-dissolving nanofiber based on poly(vinyl alcohol) and soluplus for oral cancer therapy. International Journal of Pharmaceutics. 2017;**526**(1-2):225-234. DOI: 10.1016/j.ijpharm.2017.05.004

[43] Poller B, Strachan C, Broadbent R, Walker GF. A minitablet formulation made from electrospun nanofibers.

European Journal of Pharmaceutics and Biopharmaceutics. 2017;**114**:213-220. DOI: 10.1016/j.ejpb.2017.01.022

[44] Samprasit W, Akkaramongkolporn P, Ngawhirunpat T, Rojanarata T, Kaomongkolgit R, Opanasopit P. Fast releasing oral electrospun PVP/CD nanofiber mats of taste-masked meloxicam. International Journal of Pharmaceutics. 2015;**487**(1-2):213-222. DOI: 10.1016/j.ijpharm.2015.04.044

[45] Sipos E, Szabo ZI, Redai E, Szabo P, Sebe I, Zelko R. Preparation and characterization of nanofibrous sheets for enhanced oral dissolution of nebivolol hydrochloride. Journal of Pharmaceutical and Biomedical Analysis. 2016;**109**:224-228. DOI: 10.1016/j.jpba.2016.07.004

[46] Yu DG, Shen XX, Branford-White C, White K, Zhu LM, Bligh SWA. Oral fast dissolving drug delivery membranes prepared from electrospun PVP ultrafine fibers. Nanotechnology. 2009;**20**:055104. DOI: 10.1088/0957-4484/20/5/055104

[47] Kabay G, Meydan AE, Can GK, Demirci C, Mutlu M. Controlled release of a hydrophilic drug from electrospun amyloid-like protein blend nanofibers. Materials Science and Engineering: C. 2017;**81**:271-279. DOI: 10.1016/j.msec.2017.08.003

[48] Kenawy ER, Bowlin GL, Mansfield K, Layman J, Simpson DG, Sanders EH, et al. Release of tetracycline hydrochloride from electrospun poly(ethylene-covinylacetate), poly(lactic acid), and a blend. Journal of Controlled Release. 2002;**81**(1-2):57-64. DOI: 10.1016/S0168-3659(02)00041-X

[49] Sun XZ, Williams GR, Hou XX, Zhu LM. Electrospun curcumin-loaded fibers with potential biomedica applications. Carbohydrate Polymers. 2013;**94**(1):147-153. DOI: 10.1016/j.carbpol.2012.12.064

[50] Laha A, Sharma CS, Majumdar S. Sustained drug release from multi-layered sequentially crosslinked electrospun gelatin nanofiber mesh. Materials Science and Engineering: C. 2017;**76**:782-786. DOI: 10.1016/j.msec.2017.03.110

[51] Vlachou M, Kikionis S, Siamidi A, Tragou K, Kapoti S, Ioannou E, et al. Fabrication and characterization of electrospun Nanofibers for the modified release of the Chronobiotic hormone melatonin. Current Drug Delivery. 2019;**16**(1):79-85. DOI: 10.2174/1567201815666180914095701

[52] Vlachou M, Kikionis S, Siamidi A, Tragou K, Ioannou E, Roussis V, et al. Modified in vitro release of melatonin loaded in nanofibrous electrospun mats incorporated into monolayered and three-layered tablets. Journal of Pharmaceutical Sciences. 2019;**108**(2):970-976. DOI: 10.1016/j.xphs.2018.09.035

[53] Illangakoon UE, Nazir T, Williams GR, Chatterton NP. Mebeverine-loaded electrospun nanofibers: Physicochemical characterization and dissolution studies. Pharmaceutical Nanotechnology. 2014;**103**(1):283-292. DOI: 10.1002/jps.23759

[54] Laha A, Yadav S, Majumdar S, Sharma CS. In-vitro release study of hydrophobic drug using electrospun cross-linked gelatin nanofibers. Biochemical Engineering Journal. 2016;**105**:481-488. DOI: 10.1016/j.bej.2015.11.001

[55] Hamori M, Yoshimatsu S, Hukuchi Y, Shimizu Y, Fukushima K, Sugioka N, et al. Preparation and pharmaceutical evaluation of nano-fiber matrix supported drug delivery system using the solvent-based electrospinning method. International Journal of Pharmaceutics. 2014;**464**(1-2):243-251. DOI: 10.1016/j.ijpharm.2013.12.036

[56] Hamori M, Nagano K, Kakimoto S, Naruhashi K, Kiriyama A, Nishimura A, et al. Preparation and pharmaceutical evaluation of acetaminophen nano-fiber tablets: Application of a solvent-based electrospinning method for tableting. Biomedicine & Pharmacotherapy. 2016;**78**:14-22. DOI: 10.1016/j.biopha.2015.12.023

[57] Shen X, Yu D, Zhu L, Branford-White C, White K, Chatterton NP. Electrospun diclofenac sodium loaded Eudragit® L 100-55 nanofibers for colon-targeted drug delivery. International Journal of Pharmaceutics. 2011;**408**(1-2):200-207. DOI: 10.1016/j.ijpharm.2011.01.058

[58] Wang X, Yu DG, Li XY, Bligh SWA, Williams GR. Electrospun medicated shellac nanofibers for colon-targeted drug delivery. International Journal of Pharmaceutics. 2015;**490**(1-2):384-390. DOI: 10.1016/j.ijpharm.2015.05.077

[59] Akhgari A, Heshmati Z, Afrasiabi Garekani H, Sadeghi F, Sabbagh A, Sharif Makhmalzadeh B, et al. Indomethacin electrospun nanofibers for colonic drug delivery: In vitro dissolution studies. Colloids and Surfaces B: Biointerfaces. 2017;**152**:29-35. DOI: 10.1016/j.colsurfb.2016.12.035

[60] Akhgari A, Heshmati Z, Sharif Makhmalzadeh B. Indomethacin electrospun nanofibers for colonic drug delivery: Preparation and characterization. Advanced Pharmaceutical Bulletin. 2013;**3**(1):85-90. DOI: 10.5681/apb.2013.014

[61] Akhgari A, Rotubati MH. Preparation and evaluation of electrospun nanofibers containing pectin and time-dependent polymers aimed for colonic drug delivery of celecoxib. Nanomedicine Journal. 2016;**3**(1):43-48. DOI: 10.7508/nmj.2016.01.005

[62] Bruni G, Maggi L, Tammaro L, Canobbio A, Di Lorenzo R, D'Aniello S, et al. Fabrication, physico-chemical, and pharmaceutical characterization of budesonide-loaded electrospun fibers for drug targeting to the colon. Journal of Pharmaceutical Sciences. 2015;**104**(11):3798-3803. DOI: 10.1002/jps.24587

[63] Wen P, Feng K, Yang H, Huang X, Zong MH, Lou WY, et al. Electrospun core-shell structured nanofilm as a novel colon-specific delivery system for protein. Carbohydrate Polymers. 2017;**169**:157-166. DOI: 10.1016/j.carbpol.2017.03.082

[64] Jiang J, Xie J, Ma B, Bartlett DE, Xu A, Wang CH. Mussel-inspired protein-mediated surface functionalization of electrospun nanofibers for pH-responsive drug delivery. Acta Biomaterialia. 2014;**10**(3):1324-1332. DOI: 10.1016/j.actbio.2013.11.012

[65] Yu DG, Wang X, Li XY, Chian W, Li Y, Liao YZ. Electrospun biphasic drug release polyvinylpyrrolidone/ethyl cellulose core/sheath nanofibers. Acta Biomaterialia. 2013;**9**(3):5665-5672. DOI: 10.1016/j.actbio.2012.10.021

[66] Jiang YN, Mo HY, Yu DG. Electrospun drug-loaded core-sheath PVP/zein nanofibers for biphasic drug release. International Journal of Pharmaceutics. 2012;**438**(1-2):232-239. DOI: 10.1016/j.ijpharm.2012.08.053

[67] Lee H, Xu X, Kharaghani D, Nishino M, Song KH, Lee JS, et al. Electrospun tri-layered zein/PVP-GO/zein nanofiber mats for providing biphasic drug release profiles. International Journal of Pharmaceutics. 2017;**531**(1):101-107. DOI: 10.1016/j.ijpharm.2017.08.081

[68] Bakhsheshi-Rad HR, Hadisi Z, Hamzah E, Ismail AF, Aziz M, Kashefian M. Drug delivery and cytocompatibility of ciprofloxacin loaded gelatin nanofibers-coated

Mg alloy. Materials Letters. 2017;**207**: 179-182. DOI: 10.1016/j.matlet.2017.07.072

[69] Zupančič S, Baumgartner S, Lavrič Z, Petelin M, Kristl J. Local delivery of resveratrol using polycaprolactone nanofibers for treatment of periodontal disease. Journal of Drug Delivery Science and Technology. 2015;**30**: 408-416. DOI: 10.1016/j.jddst.2015.07.009

[70] Sultanova Z, Kaleli G, Kabay G, Mutlu M. Controlled release of a hydrophilic drug from coaxially electrospun polycaprolactone nanofibers. International Journal of Pharmaceutics. 2016;**505**(1-2):133-138. DOI: 10.1016/j.ijpharm.2016.03.032

[71] Paaver U, Heinamaki J, Laidmae I, Lust A, Kozlova J, Sillaste E, et al. Electrospun nanofibers as a potential controlled-release solid dispersion system for poorly water-soluble drugs. International Journal of Pharmaceutics. 2015;**479**(1):252-260. DOI: 10.1016/j.ijpharm.2014.12.024

[72] Karthikeyan K, Guhathakarta S, Rajaram R, Korrapati PS. Electrospun zein/eudragit nanofibers based dual drug delivery system for the simultaneous delivery of aceclofenac and pantoprazole. International Journal of Pharmaceutics. 2012;**438**(1-2): 117-122. DOI: 10.1016/j.ijpharm.2012.07.075

Chapter 3

Preparation, Characterization, and Applications of Electrospun Carbon Nanofibers and Its Composites

Mayakrishnan Gopiraman and Ick Soo Kim

Abstract

Carbon nanofibers (CNFs) and its composites have gained vast attention due to its exceptional chemical and textural properties. So far, various multifunctional carbon nanofibers and its composites are developed with highly unique and tunable morphology. In this chapter, we reviewed unique fabrication methods that are recently reported and its characterization techniques such as SEM, FE-SEM, TEM, WAXD, XPS, AFM, and Raman. In addition, catalytic, energy, and environmental applications of carbon nanofiber composites (metals and/or metal oxide nanoparticles incorporated and/or decorated hybrid carbon nanofibers) are discussed. Preparation and characterization of electrospun carbon nanofiber composites and its applications in catalysis and energy storage are the main focus of this chapter.

Keywords: electrospinning, carbon nanofibers, hollow structures, composites, catalysis

1. Introduction

Carbon nanofibers have received growing interests due to their unique chemical and physical properties, depending upon their size, surface area, and shape [1, 2]. Indeed the attractive structural, electrical, and mechanical properties of carbon nanotubes (CNTs) make it an ideal supporting material for various applications. Particularly, the CNTs can be used as an efficient support for the decoration of catalytic active materials. Carbon nanofibers (CNFs) also have the similar physicochemical properties to CNTs and the diameter varying from some tens of nanometers to 500 nm [3] and are also suitable to be used as catalyst support. Electrospinning is a very simple but powerful method for the fabrication of high-quality carbon nanofibers [4]. In general, the CNFs with sub-micrometer diameters as well as some tens of nanometers to 500 nm are prepared by carbonization of electrospun polymer nanofibers under inert atmosphere at high temperature [5]. Undoubtedly, polyacrylonitrile (PAN) is a well-known and efficient precursor for the fabrication of carbon fibers [6]; therefore, several attempts were made to prepare the electrospun-derived carbon nanofibers from PAN [7]. Several approaches, including wet-chemical synthesis [8, 9], electrodeposition [10, 11], and dry synthesis [12–20], are developed to obtain various multifunctional active materials loaded with carbon nanocomposites. By using these techniques, various types of metal or metal oxide nanoparticles (NPs), such as Au,

Co, Ru, Pt, Pd, Ag, Co, Rh, Ti and Cu, have been decorated or immobilized on/into the carbon nanofibers. These metal NP-supported CNF nanocomposites have shown great promises in catalysis [21], fuel cells [22], and highly sensitive chemical/biological sensing applications [23, 24]. In particularly, the CNF composites showed excellent results in various catalytic systems such as in photocatalytic activity [25, 26], water gas shift reactions (WGS) [27], enzyme immobilization or biocatalysts [28, 29], and direct oxidation of alcohols [30]. So far, TiO_2-deposited CNFs have gained much attention in the photocatalytic reactions, and a considerable number of reports are available in the literature. Alike, Pd NP-supported CNFs are often preferred for the catalytic organic conversions such as hydrogenation reaction [31] and Heck coupling reaction [8]. It is proven that the CNFs are one of the highly suitable supports for the decoration of Pd NPs and the resultant Pd/CNF composite often demonstrated an enhanced catalytic activity [32]. In fact, the unique structure, high conductivity, huge surface area, and chemical inertness of CNFs often help to obtain high dispersion of metal nanoparticles on CNFs. Most of the Pd NP-supported CNFs showed better activity than the conventional Pd/C catalysts [8, 33].

In this chapter, we discussed the preparation and characterization of electrospun carbon fibers and its composites. The contributions of CNF composites in various catalytic systems (such as photocatalytic activity, water gas shift reactions (WGS), enzyme immobilization or biocatalysts, and direct oxidation of alcohols) are also discussed in detail.

2. Preparation and characterization of carbon nanofiber composites

2.1 Electrospun carbon nanofibers

Electrospinning is a straightforward method to obtain the nanofibers. **Figure 1** shows the fundamental electrospinning setup and the list of important parameters to be controlled. The nanofibers can be produced by applying high voltage to a polymer solution which could create electrostatically repulsive force and an electric field between two electrodes, so that the nanofibers can be formed [34]. Obviously, the formation of nanofibers is highly dependent on the viscosity and electric conductivity of the polymeric fluids, humidity, and applied voltage [35]. To date,

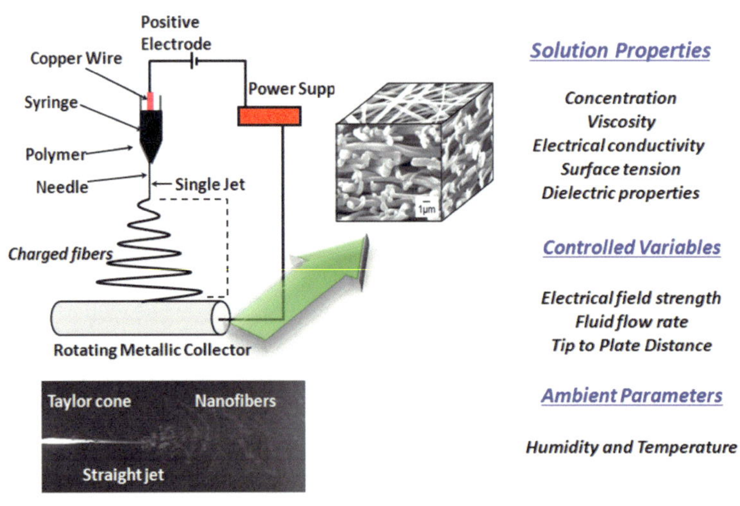

Figure 1.
Scheme of fundamental setup for electrospinning and electrospinning parameters.

over 100 kinds of polymers have been employed to produce their nanofibers via electrospinning. However, a very limited number of polymers such as polyacrylonitrile (PAN), polyimide (PI), poly(vinyl alcohol) (PVA), poly(vinylidene fluoride) (PVDF), cellulose acetate, and pitch have been successfully used to obtain carbon nanofibers [36]. The carbon nanofibers are often characterized by various techniques. Scanning electron microscope (SEM) is one of the very common methods to characterize the CNFs. The surface morphology, particularly the fiber diameter, uniformity, and surface smoothness are often studied by SEM analysis. Alike, transmission electron microscopy (TEM) and atomic force microscopy (AFM) are also employed for the detail surface analysis. The crystalline and amorphous nature of the CNFs is often investigated by means of X-ray diffraction (XRD) analysis. Raman spectroscope is a very useful technique for the analysis of G-band and D-band of CNFs. X-ray photoemission spectroscopy (XPS) was also effectively used to analyze the CNFs. The specific surface area and textural properties such as pore volume and pore size of CNFs are evaluated by using the Brunauer-Emmett-Teller (BET) method.

Kim et al. [37] prepared CNFs via electrospinning by using PVA and DMF as precursor and solvent, respectively. In a typical preparation method, PVA was dissolved in DMF and the polymer mixture was electrospun. In the first step, the resultant nanofiber webs were oxidatively stabilized at 280°C under air flow (heating at 1°C/min). Then the stabilized nanofiber web was activated by steam resulting in activated carbon nanofibers. The stabilized nanofiber webs were heated at a rate of 5°C/min up to 700, 750, and 800°C and activated for 30 min by supplying 30 vol.% of steam in a carrier gas of N_2. They confirmed that the resultant CNFs have well-developed mesopores and the CNFs demonstrated excellent specific capacitance (173 F/g at 10 mA/g).

Kuzmenko and co-workers [38] prepared nitrogen-doped carbon nanofibrous mats from regenerated cellulose impregnated with ammonium chloride. The ammonium chloride provided the thermal stabilization of incompletely regenerated cellulose fibers. In a typical preparation, cellulose acetate solution was prepared in acetone/DMAc mixture which was subsequently electrospun. The voltage was 25 kV, distance between needle and collector was 25 cm, and the process was performed at temperature around 20 ± 2°C and relative humidity 45–60%. Aluminum foil was used for collecting the nanofibers. The prepared cellulose acetate nanofibers were deacetylated by using dilute NaOH solution. Then the regenerated cellulose webs were impregnated with NH_4Cl by immersion in 0.3 M aqueous solution of NH_4Cl for 24 h at 20 ± 2°C. The NH_4Cl-treated regenerated cellulose samples were carbonized in a quartz tube furnace for general annealing in N_2 flow (1 L/min) by heating up to 800°C with the heating rate of 5°C/min. **Figure 2** shows the SEM images of the CNFs synthesized from the regenerated cellulose [38].

2.2 Porous carbon nanofibers with hollow cores

Kim et al. [39] successfully prepared porous CNFs with hollow cores through the thermal treatment of electrospun copolymeric nanofiber webs. **Figure 3** shows the schematic diagram for producing porous CNFs with hollow cores. For the preparation of pores CNFs with hollow cores, PAN and poly(methyl methacrylate) (PMMA) polymers were chosen. The PAN is a widely used precursor for the preparation of CNFs, and the PMMA can be thermally decomposed at elevated temperatures. Dissolving these two polymers (PAN and PMMA) in a solvent would create phase separation [continuous phase (sea) changes into pore walls (or skeletons of nanofibers) and the discontinuous phase (islands) changes into many hollow pores], which results in the *sea-islands* feature. It is well known that

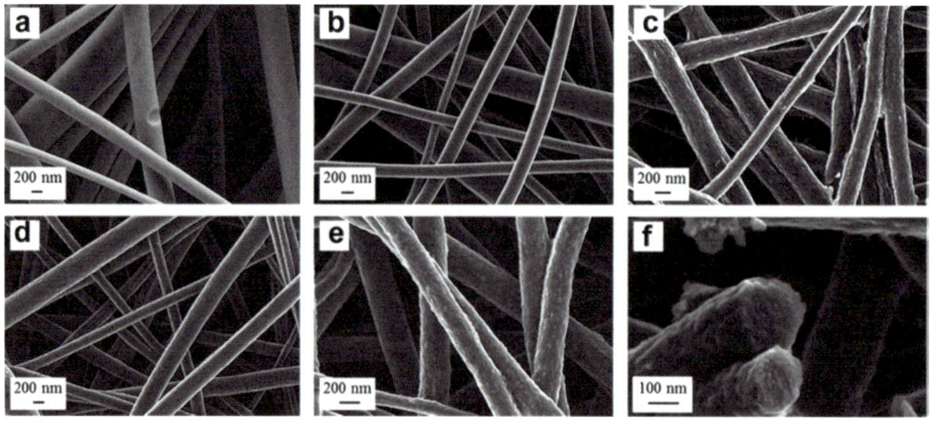

Figure 2.
(a-f) SEM images of the CNFs synthesized from the differently regenerated cellulose with additional NH_4Cl impregnation [38].

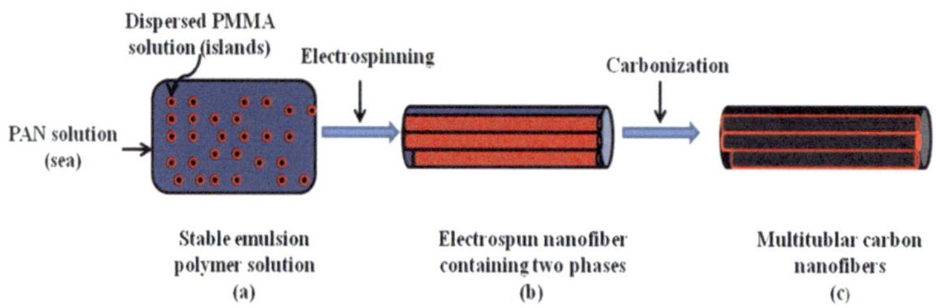

Figure 3.
Schematic diagram for the preparation of porous CNFs with hollow cores. (a) Preparation of stable polymer solutions from two separate phases; nanoscale phase separation occurs due to their different molecular weights; PMMA forms the discontinuous phase and PAN forms the continuous phase; (b) nanofiber formation (with two phases) by electrospinning; (c) removal of the PMMA phase at elevated temperatures [39].

the low-surface-tension polymer (PAN) would occupy the continuous phase of the solution (*sea*), while the high-surface-tension polymer (PMMA) forms the discontinuous phase (*islands*). In fact the two separate phases are due to the intrinsic properties (e.g., interfacial tension, viscosity, elasticity) of the polymers [40]. The electrospinning technique was used to obtain the PAN/PMMA nanofibers containing two separate phases. The thermal treatment of PAN/PAMM nanofibers at over 1000°C would eventually form the porous carbon nanofibers with hollow cores. The complete removal of PMMA phase and the transformation continuous PAN phase would result in the formation of porous CNFs with hollow cores (**Figure 4**).

Highly flexible N- and O-containing porous ultrafine CNFs were prepared by Wei and co-workers [41]. The ultrafine porous CNFs were obtained by simply varying the PAN/PMMA ratios (10/0, 7:3, 5:5, and 3:7). Briefly, PAN/PMMA solutions with different ratios (10:0, 7:3, 5:5, and 3:7) are prepared in DMF. For better dispersion, the PAN/PMMA solution was sonicated followed by stirring at 60°C for 2 h. The polymer blend was electrospun under an electric field of 9 kV at a tip-to-collector distance of 15 cm. The resultant electrospun PAN/PMMA nanofibers were stabilized under air flow at 300°C with the heating rate of 1°C/min for 1 h. Subsequently, the stabilized nanofibers were carbonized under N_2 atmosphere at 900°C with heating rate of 5°C/min for 1 h. It was proven that the increasing the ratio of PMMA would result in the formation of ultrafine CNFs. **Figure 5** shows the FE-SEM images of CNFs; CNFs, 7:3; CNFs, 5:5; and *u*-CNFs, 7:3. The FE-SEM images show that the morphology of CNFs is homogeneous, continuous, and a typical cylindrical shape.

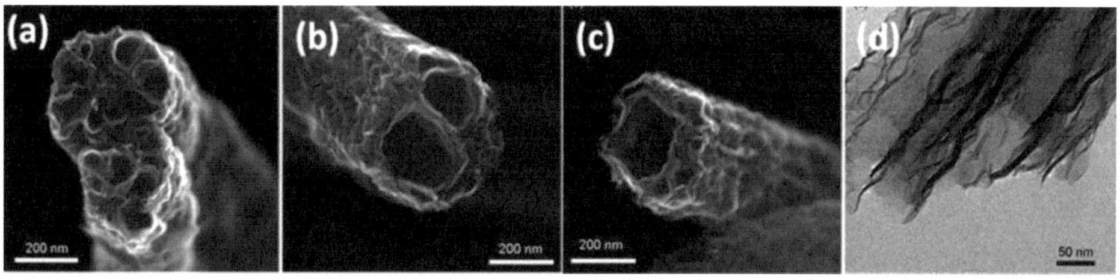

Figure 4.
(a–c) Cross-sectional TEM images of CNFs thermally treated at 2800°C [PAN:PMMA (a) 5:5, (b) 7:3, and (c) 9:1] and (d) TEM image of CNFs showing structurally developed core walls after thermal treatment [39].

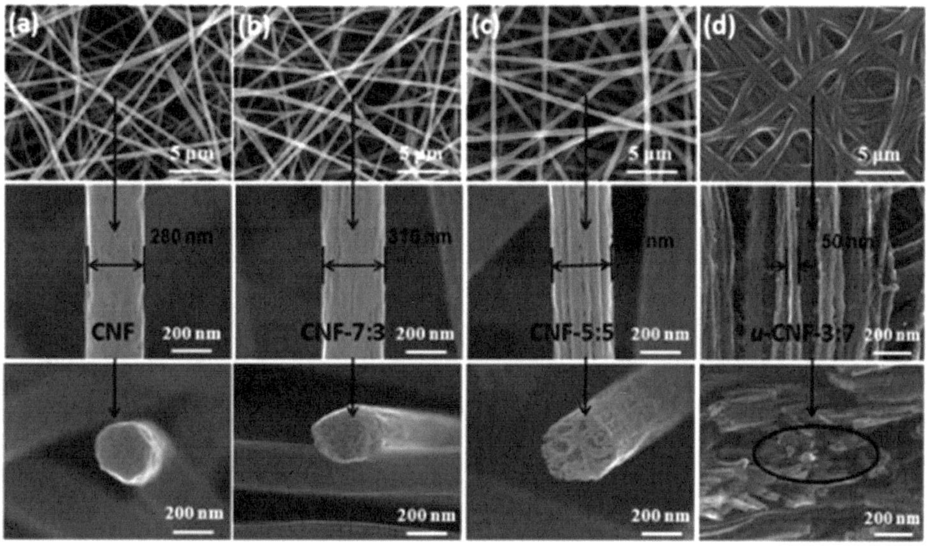

Figure 5.
FE-SEM images of electrospun PAN/PMMA nanofibers with different mixing ratios. (a) CNFs; (b) PAN/PMMA, 7:3; (c) PAN/PMMA, 5:5; and (d) PAN/PMMA, 3:7 [41].

The FE-SEM images of CNFs, 7:3, and CNFs, 5:5, showed that the CNFs have several hollow cores along the fiber axis. Notably, the morphology of the u-CNFs (3:7) was completely changed. The morphology of CNFs (3:7) fibers was homogenous, continuous, and a cylindrical shape with an average diameter of ~50 nm. In fact, the complete decomposition of PMMA during the thermal treatment is the main reason. The BET specific surface area of the u-CNFs-3,7 was determined to be 467.57 m^2/g and pore volume of 1.15 cm^3 g^{-1} and an average pore size of 9.48 nm.

Chang et al. [42] introduced a novel technique of centrifuged-electrospinning for the preparation of ultrathin carbon fibers. **Figure 6** shows the preparation diagram of the ultrathin porous CNFs by centrifuged-electrospinning. In a typical procedure, PAN/PMMA polymer blend was prepared in DMF at different weight ratios of 80/20 (PAN80/PMMA20) and 10/90 (PAN10/PMMA90). The polymer blends were used for the preparation of PAN/PMMA nanofibers by centrifuged electrospinning. The centrifuged-electrospinning conditions were as follows: an applied positive voltage of 45 kV, a three-phase induction motor spinning at 4000 rpm, a syringe feed rate of 1.5 mL/min, and a stainless steel ring with a diameter of 50 cm as the collector. Finally, the resultant PAN/PMMA nanofibers were stabilized at 280°C for 2 h at a heating rate 0.5°C/min in air atmosphere and then carbonized at 800°C for 4 h under argon atmosphere at a heating rate of 5°C/min. **Figure 6** (b, c, and d) shows the TEM images of ultrathin micro-/mesoporous CNFs.

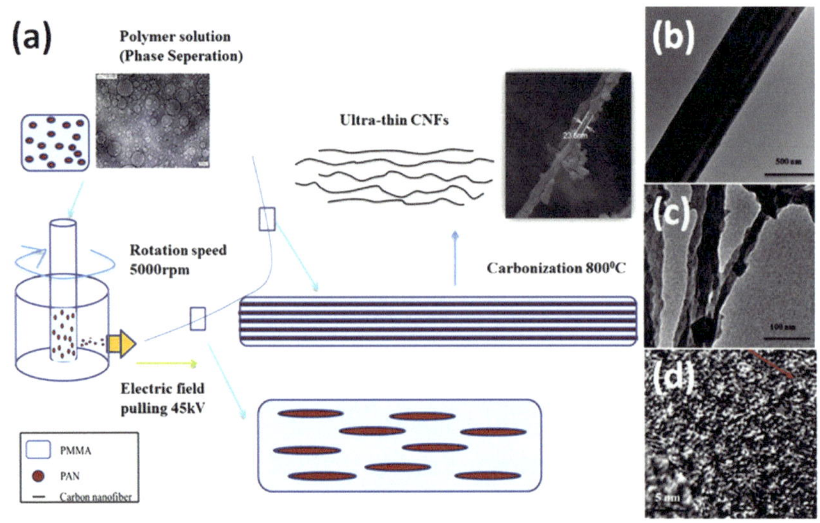

Figure 6.
(a) Schematic diagram for the preparation of ultrathin micro-/mesoporous CNFs by centrifuged-electrospinning followed by carbonization and (b, c, and d) TEM images of ultrathin micro-/mesoporous CNFs [42].

2.3 Carbon nanofiber composites

Recently, preparation of metal oxide-supported carbon nanofiber composites via electrospinning has been extensively studied. The carbon nanocomposites are used in various applications such as energy conversion and storage, capacitive deionization, catalysis, adsorption/separation, and in the field of biomedicine. In order to achieve higher activity, various synthetic routes were developed to achieve porous carbon nanofibers composites with high surface area and tunable pore size distribution. Most of the preparation methods involve carbonization process at elevated temperatures of typically above 1200°C. So far, various metal or metal oxide nanoparticle (Pd, Pt, Ti, Ag, Au, Cu, Ni, Zn, and Ru)-supported CNF nanocomposites were reported [21, 43].

Atchison et al. [44] prepared metal carbide-supported carbon nanocomposites through carbothermal reduction process. Zirconium carbide/carbon nanocomposite (ZrC/C), titanium carbide/carbon nanocomposite (TiC/C), and niobium carbide/carbon nanocomposite (NbC/C) were prepared by electrospinning followed by carbothermal reduction at elevated temperatures. Cellulose acetate and PVP were used as precursor.

Chen et al. [45] prepared Pd nanoparticle-supported carbon nanofibers (Pd-NP/CEPFs: Pd-NP/CEPFs) through the electrospinning process. Shortly, electrospinning solution was prepared by using 10 wt% PAN and 3.3 wt% Pd(OAc)$_2$ in DMF. The electrospinning process was performed in an electric field of 30 kV and the tip-to-collector distance of 30 cm. Then the electrospun PAN/Pd(OAc)$_2$ nanofiber involved three steps as follows: (1) 210°C annealing for 1 h under air flow for the oxidation of PAN, (2) heating up to 400°C at a rate of 5°C/min and annealing for 2 h in H_2 and Ar mixture (H_2/Ar = 1/3) atmosphere for the reduction of Pd^{2+}, and (3) heating up to 550°C at a rate of 5°C/min and annealing for 1 h in Ar for the formation of metal nanoparticles on/in the carbonized nanofibers (**Figure 7**).

Zhang et al. [46] obtained AgNP-immobilized carbon nanocomposite by a two-step preparation: electrospinning followed by the hydrothermal growth of the AgNPs on the CNFs (**Figure 8**). In a typical procedure, the electrospinning solution

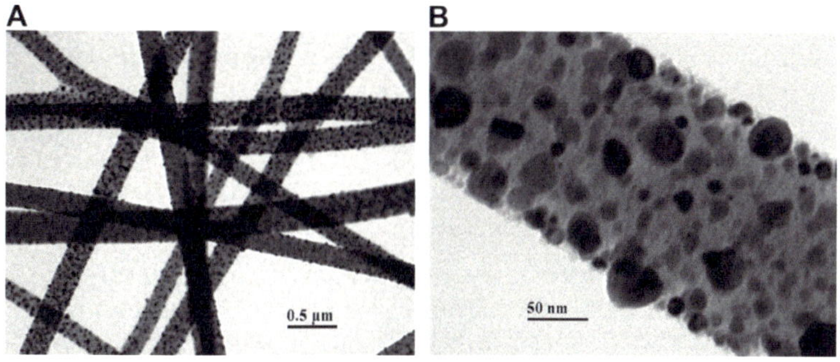

Figure 7.
TEM images of Pd-NP/CENFs. (A) Lower magnification; (B) higher magnification [45].

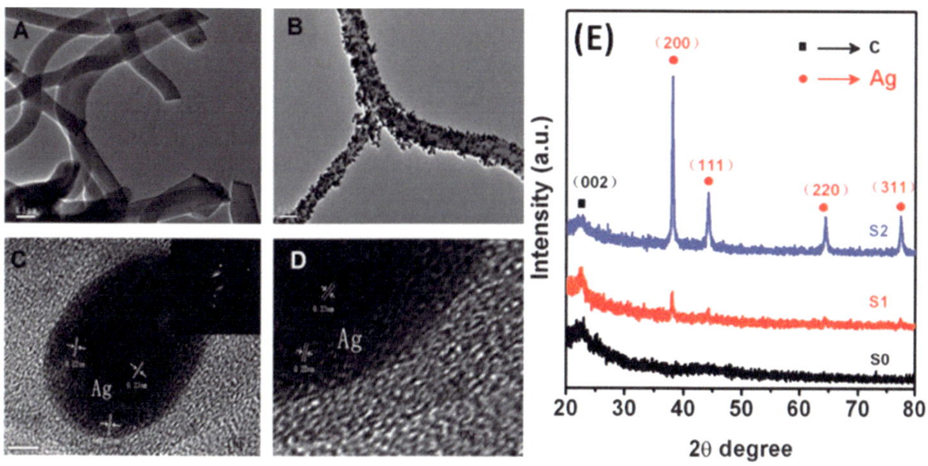

Figure 8.
TEM images of sample (A) CNFs and (B) CNFs/AgNPs, (C) HRTEM images of CNFs/AgNPs, and (E) XRD patterns of CNFs/AgNPs and CNFs.

of PAN was prepared in DMF, and it was electrospun at an applied electric voltage of 10 kV. The PAN nanofibers were then stabilized under air flow at 270°C for 1 h and subsequently carbonized under N_2 atmosphere at 1000°C for 1 h at the rate of 5°C/min. After the preparation of CNFs from PAN nanofibers, the CNFs were treated with HNO_3, centrifuged, and washed with water for several times. Finally an aqueous mixture of glucose, $Ag(NH_3)_2OH$, and CNFs was stirred for 5 min. After being stirred for more than 5 min, the mixture was transferred into a Teflon-lined autoclave, and it was sealed in a stainless steel tank and heated at 180°C for 3 h. Finally the AgNP-immobilized CNFs (Ag/CNFs) were obtained.

Yu and co-workers [47] prepared electrospun Ag/g-C_3N_4-loaded composite carbon nanofibers (Ag/g-C_3N_4/CNFs) through combing the electrospinning technology and carbonization treatment. The microstructure of Ag/g-C_3N_4/CNFs was characterized by XRD, FE-SEM, EDS, TEM, and XPS.

Ghouri et al. [48] achieved Co/CeO_2-decorated carbon nanofibers (Co/CeO_2/CNFs) by calcination of electrospun nanofibers composed of cerium (III) acetate hydrate, cobalt (II) acetate tetrahydrate, and poly(vinyl alcohol) in nitrogen environment at 700°C. PVA was used as carbon source due to its high carbon content. In a typical preparation, CoAc and CeAc aqueous solutions were prepared in distilled water. The resultant aqueous solutions were mixed with PVA aqueous solution. After stirring for 6 h, the mixture was electrospun at high voltage of 22 kV using DC power supply at room temperature with 65% relative humidity. The tip-to-collector distance of 22 cm was fixed. Finally, the dried nanofiber mats were calcined at 700°C for 6 h in N_2 flow with a heating rate of 2.0°C/min. The physicochemical properties of the Co/CeO_2/CNFs were characterized by XRD, FE-SEM, EDS, TEM, XPS, and Raman.

The utilization of noble metals in green technologies has garnered an increasing level of research interest. Particularly, the Pt-based nanocomposites are often preferred as the anode because of their excellent performance in catalyzing the dehydrogenation of methanol. For example, Formo et al. [49] achieved Pt nanostructure-supported CNF nanofibers through electrospinning followed by calcination in air at 510°C for 6 h.

3. Applications of carbon nanofiber composites

Electrospun carbon nanofibers have proven to be efficient catalytic supports owing to the high porosity and large surface areas. The high porosity in a nonwoven mat of nanofibers enables direct growth of catalytic nanostructures. Till date, there are number of applications found for the electrospun carbon nanofibers and its composites.

3.1 Carbon nanocomposites in organic transformations

Owing to high surface area, porosity, stability, metal-support interaction, smaller particle size, and high dispersion in reaction medium, the metal nanoparticle-supported carbon nanocomposites demonstrated excellent activity in organic reactions. They can be highly reusable due its stability which is one of the hallmarks of the carbon nanocomposites.

Palladium-catalyzed Sonogashira coupling reaction is the most straightforward and powerful method used for the construction of C(sp_2)–C(sp) bond, drugs, and polymeric materials [50]. The conventional protocols of the Sonogashira reactions are carried out in the homogeneous phase, using soluble palladium (Pd) composites such as $Pd(PPh_3)_4$, $Pd(PPh_3)_2Cl_2$, and $Pd(OAc)_2$ as catalysts in the presence of CuI as co-catalyst. Even with the high reaction rate and high turnover numbers, homogeneous catalysis has a number of disadvantages, in particular the lack of reuse of the catalyst. Chen et al. [45] developed Pd-supported CNF catalytic system for the Sonogashira reaction. **Figure 9** shows Pd-NP/CENF catalyzed Sonogashira reaction of iodobenzene and phenylacetylene in liquid phase. The catalyst showed superior catalytic activity toward the Sonogashira reaction. In addition, the catalyst was found to be highly reusable, at least for 10 runs without any significant loss in its catalytic activity.

Alike, electrospun Ag/g-C_3N_4-loaded composite carbon nanofibers (Ag/g-C_3N_4/CNFs) were used for the conversion of 4-nitrophenol to 4-aminophenol and benzylamine to N-benzylbenzaldimine [44]. The Ag/g-C_3N_4/CNFs offered the significant advantages, such as low catalyst use, high activity, easy recycling, and excellent stability. In fact, the synergistic effect between catalytic activity of Ag nanoparticles

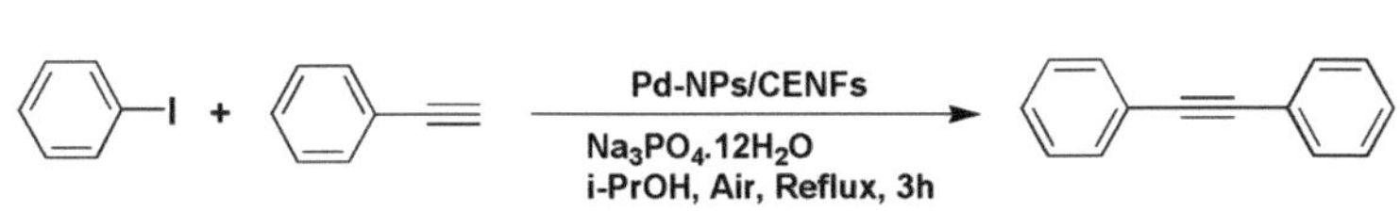

Figure 9.
The Sonogashira reaction equation of iodobenzene and phenylacetylene in liquid phase catalyzed by Pd-NP-supported CNFs [45].

(NPs) and g-C_3N_4 and excellent adsorption capacity of carbon nanofibers are the main reason for the excellent catalytic activity.

3.2 Carbon nanocomposites in catalytic reduction of 4-nitrophenol

Catalytic transformation of 4-nitrophenol to 4-aminophenol is one of the very significant reactions in green chemistry [51]. It is well known that the nitrophenols are harmful to the environment, and therefore the US Environmental Protection Agency has listed it as 114th organic pollutant [52]. In recent days, catalytic conversion of nitrophenols to aminophenols is widely studied. In fact, the catalytic product, aminophenols, can be used as an excellent intermediate in synthesizing various drugs and reducing agent. The 4-nitrophenols are often employed as a photographic developer, corrosion inhibitor, anticorrosion lubricant, and hair-dyeing agent [53]. Although the reaction is very simple, greener, and most efficient, the reaction without metal catalysts is not achievable. To perform this reaction, various metal catalysts (based on graphene oxide, silica, alumina, activated carbon, CNTs, fullerenes, and so on) are developed and proven to be an excellent candidate for the reduction of 4-nitrophenols to 4-aminophenol. For example, RGO-$ZnWO_4$-Fe_3O_4, AgNPs-rGO, PdNiP/RGO, and NiNPs/silica are reported for the reduction of nitrophenol [20].

Carbon nanofibers/silver nanoparticle (CNFs/AgNPs) composite nanofibers were used for the reduction of 4-nitrophenol (4-NP) with $NaBH_4$ [46]. The reaction was tracked by time-dependent UV-visible spectroscopy (**Figure 10**). It was found that the CNF/AgNP composite demonstrated an excellent catalytic activity in the reduction of 4-nitrophenol. The catalytic efficiency was found to be enhanced with the increasing of the content of silver on the CNF/AgNP catalyst. Reaction kinetic was studied for the CNF/AgNP reduction of 4-nitrophenol. It was reported that the rate constant of $6.2 \times 10^{-3}\ s^{-1}$ was determined for the reduction of 4-nitrophenol over CNF/AgNP catalyst. The excellent active might be attributed to the high surface areas of Ag NPs and synergistic effect on delivery of electrons between CNFs and AgNPs. The CNF/AgNP composite nanofibers can be easily recycled and reused without any significant loss in its catalytic activity.

3.3 Carbon nanofiber composites in energy applications

Due to the excellent properties such as high surface area, conductivity, and porosity, the CNF-based nanocomposites are widely used for the energy applications. Without a doubt, the development of electrochemical energy storage systems (EES)

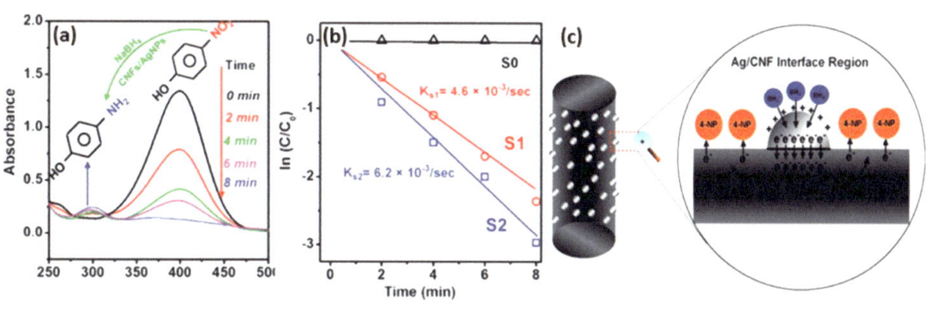

Figure 10.
Catalytic evolution of CNFs/AgNPs. (a) UV-vis absorption spectra during the reduction of 4-nitrophenol over CNFs/AgNPs; (b) ln(C/C₀) and C/C₀ vs. reaction time for the reduction of 4-nitrophenol, S0 = fresh CNFs, S1 = CNFs/AgNPs, and S2 = CNFs/AgNPs; (c) proposed mechanism of the catalytic reduction of 4-nitrophenol with the CNF/AgNP composite nanofibers [46].

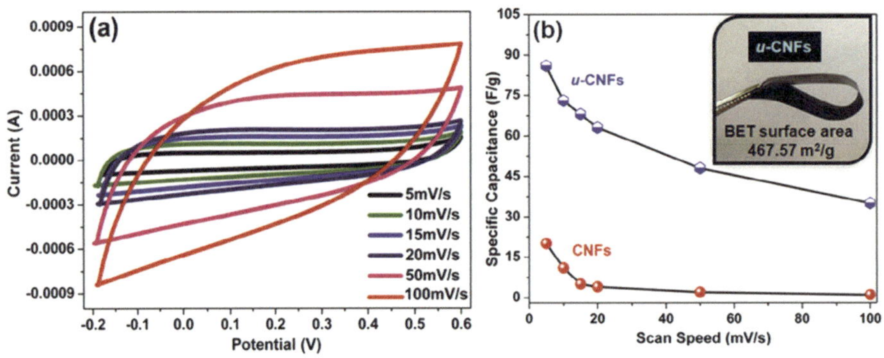

Figure 11.
(a) Cyclic voltammetry results for the ultrafine CNFs at different scan rates in 1.0 mol/L H_2SO_4 and; (b) specific capacitance for the u-CNFs (3:7) and CNFs at different scan rates in 1.0 mol/L H_2SO_4.

is largely focused due to its vital demand for clean and sustainable energy. Mainly three types of devices are very important and most commercialized energy storage systems such as batteries, electrochemical capacitors (ECs), and fuel cells [54].

The ultrafine CNFs prepared via electrospinning of PAN/PMMA blend followed by thermal treatment in inert atmosphere were used as a flexible electrode material for the supercapacitor applications [41]. **Figure 11** shows the cyclic voltammetry results for the ultrafine CNFs at different scan rates in 1.0 mol/L H_2SO_4 and specific capacitance for the ultrafine CNFs at different scan rates in 1.0 mol/L H_2SO_4. The ultrafine CNFs demonstrated an enhanced specific capacitance of 86 F g^{-1} in 1 mol/L H_2SO_4. Being a flexible electrode material, this is the highest specific capacitance for the CNFs reported so far. The excellent specific capacitance of ultrafine CNFs is due to its unique properties. The results proved that the fiber diameter of ultrafine CNFs was about 50 nm. The XPS and Raman studies confirmed the presence of N and O in the form of various functional groups such as pyridinic, benzenoid amine, graphitic N, and N-oxides. High specific surface area of 467.57 m^2/g with an excellent pore volume (1.15 cm^3 g^{-1}) and pore size (9.48 nm) was determined for the ultrafine CNFs. The BET results confirmed the interconnected micro-/meso-/macropores on the surface of the ultrafine CNFs.

The Pt nanostructure-supported CNF nanocomposite was employed for the direct oxidation of methanol [49]. It was found that the Pt nanostructure-supported CNF nanocomposite showed better activity than the commercial Pt/C which may be due to the synergistic effect of the underlying anatase surface and the Pt nanostructures with well-defined facets. Alike, Co/CeO_2-decorated carbon nanofibers were developed for the methanol oxidation [48]. The results showed that the electrocatalytic activity of the Co/CeO_2-decorated carbon nanofibers toward methanol oxidation was excellent. Interestingly, the introduced catalyst revealed negative onset potential (50 mV vs. Ag/AgCl) which is a superior value among the reported non-precious electrocatalyst.

4. Conclusion

Electrospinning is one of the simple and effective techniques for the fabrication of carbon nanofiber. Certainly, metal-supported carbon nanofibers demonstrated excellent activity in various applications such as catalysis, energy, and environmental. In this chapter, we have summarized the recent progress in the research on the preparation methods, characterization, and applications of electrospun carbon nanofibers and its composites.

Author details

Mayakrishnan Gopiraman[1] and Ick Soo Kim[2]*

1 Department of Applied Bioscience, College of Life and Environmental Science, Konkuk University, Seoul, South Korea

2 Nano Fusion Technology Research Group, Division of Frontier Fibers, Institute for Fiber Engineering (IFES), Interdisciplinary Cluster for Cutting Edge Research (ICCER), Shinshu University, Ueda, Japan

*Address all correspondence to: kimicksoo@hotmail.com

References

[1] Burda C, Chen XB, Narayanan R, El-Sayed MA. Chemistry and properties of nanocrystals of different shapes. Chemical Reviews. 2005;**105**:1025-1102. DOI: 10.1021/cr030063a

[2] De Jong KP, Geus JW. Carbon nanofibers: Catalytic synthesis and applications. Catalysis Reviews. 2000;**42**:481-510. DOI: 10.1081/CR-100101954

[3] Vamvakaki V, Tsagaraki K, Chaniotakis N. Carbon nanofiber-based glucose biosensor. Analytical Chemistry. 2006;**78**:5538-5542. DOI: 10.1021/ac060551t

[4] Kim C, Yang KS, Kojima M, Yoshida K, Kim YJ, Kim YA, et al. Fabrication of electrospinning-derived carbon nanofiber webs for the anode material of lithium-ion secondary batteries. Advanced Functional Materials. 2006;**16**:2393-2397. DOI: 10.1002/adfm.200500911

[5] Huang CB, Chen SL, Reneker DH, Lai CL, Hou HQ. High-strength mats from electrospun poly(p-phenylene biphenyltetracarboximide) nanofibers. Advanced Materials. 2006;**18**:668-671. DOI: 10.1002/adma.200501806

[6] Donnet JB, Bansal RC. Carbon Fibers. New York: Marcel Dekker; 1990

[7] Wang Y, Santiago-Aviles JJ. Conductivity measurement of electrospun PAN-based carbon nanofiber. Journal of Materials Science Letters. 2002;**21**:1055-1057

[8] Yoon B, Wai CM. Microemulsion-templated synthesis of carbon nanotube-supported Pd and Rh nanoparticles for catalytic applications. Journal of the American Chemical Society. 2005;**127**(49):17174-17175. DOI: 10.1021/ja055530f

[9] Selvamani A, Babu CM, Ramkumar V, Sundaravel B. Reduced graphene oxide decorated Au nanoparticles as an efficient electrode for the determination of hydroquinone. Nano Progress. 2019;**1**:9-14

[10] Day TM, Unwin PR, Macpherson JV. Factors controlling the electrodeposition of metal nanoparticles on pristine single walled carbon nanotubes. Nano Letters. 2007;7:51-57. DOI: 10.1021/nl061974d

[11] Lin Z, Ji L, Zhang X. Electrodeposition of platinum nanoparticles onto carbon nanofibers for electrocatalytic oxidation of methanol. Materials Letters. 2009;**63**:2115-2118. DOI: 10.1016/j.matlet.2009.07.005

[12] Yuan G, Gopiraman M, Cha HJ, Soo HD, Chung IM, Kim IS. Interconnected ruthenium dioxide nanoparticles anchored on graphite oxide: Highly efficient candidate for solvent-free oxidative synthesis of imines. Journal of Industrial and Engineering Chemistry. 2017;**46**:279-288. DOI: 10.1016/j.jiec.2016.10.040

[13] Gopiraman M, Deng D, Babu SG, Hayashi T, Karvembu R, Kim IS. Sustainable and versatile CuO/GNS nanocatalyst for highly efficient base free coupling reactions. ACS Sustainable Chemical Engineering. 2015;**3**:2478-2488. DOI: 10.1021/acssuschemeng.5b00542

[14] Gopiraman M, Babu SG, Khatri Z, Kai W, Kim YA, Endo M, et al. Dry synthesis of easily tunable nano ruthenium supported on graphene: Novel nanocatalysts for aerial oxidation of alcohols and transfer hydrogenation of ketones. The Journal of Physical Chemistry C. 2013;**117**:23582-23596. DOI: 10.1021/jp402978q

[15] Gopiraman M, Babu SG, Khatri Z, Kai W, Kim YA, Endo M, et al. An efficient, reusable copper-oxide/carbon-nanotube catalyst for N-arylation of imidazole. Carbon. 2013;**62**:135-148. DOI: 10.1016/j.carbon.2013.06.005

[16] Gopiraman M, Babu SG, Karvembu R, Kim IS. Nanostructured RuO_2 on MWCNTs: Efficient catalyst for transfer hydrogenation of carbonyl compounds and aerial oxidation of alcohols. Applied Catalysis A: General. 2014;**484**:84-96. DOI: 10.1016/j.apcata.2014.06.032

[17] Gopiraman M, Bang H, Babu SG, Wei K, Karvembu R, Kim IS. Catalytic N-oxidation of tertiary amines on RuO_2 NPs anchored graphene nanoplatelets. Catalytic Science and Technology. 2014;**4**:2099-2106. DOI: 10.1039/C3CY00963G

[18] Gopiraman M, Karvembu R, Kim IS. Highly active, selective, and reusable RuO_2/SWCNT catalyst for Heck olefination of aryl halides. ACS Catalysis. 2014;**4**:2118-2129. DOI: 10.1021/cs500460m

[19] Gopiraman M, Babu SG, Khatri Z, Kai W, Kim YA, Endo M, et al. Facile and homogeneous decoration of RuO_2 nanorods on graphene nanoplatelets for transfer hydrogenation of carbonyl compounds. Catalysis Science and Technology. 2013;**3**:1485-1489. DOI: 10.1039/C3CY20735H

[20] Saravanamoorthy S, Chung IM, Ramkumar V, Ramaganth B, Gopiraman M. Highly active and reducing agent-free preparation of cost-effective NiO-based carbon nanocomposite and its application in reduction reactions under mild conditions. Journal of Industrial and Engineering Chemistry. 2018;**60**:91-101. DOI: 10.1016/j.jiec.2017.10.006

[21] van der Lee MK, van Dillen J, Bitter JH, de Jong KP. Deposition precipitation for the preparation of carbon nanofiber supported nickel catalysts. Journal of the American Chemical Society. 2005;**127**:13573-13582. DOI: 10.1021/ja053038q

[22] Cao L, Scheiba F, Roth C, Schweiger F, Cremers C, Stimming U, et al. Novel nanocomposite Pt/RuO_2 × H_2O/carbon nanotube catalysts for direct methanol fuel cells. Angewandte Chemie International Edition. 2006;**45**:5315-5319. DOI: 10.1002/anie.200601301

[23] Hrapovic S, Liu Y, Male KB, Luong JH. Electrochemical biosensing platforms using platinum nanoparticles and carbon nanotubes. Analytical Chemistry. 2004;**76**:1083-1088. DOI: 10.1021/ac035143t

[24] Yang M, Yang Y, Liu Y, Shen G, Yu R. Platinum nanoparticles-doped sol–gel/carbon nanotubes composite electrochemical sensors and biosensors. Biosensors and Bioelectronics. 2006;**21**:1125-1131. DOI: 10.1016/j.bios.2005.04.009

[25] Zhang Z, Shao C, Li X, Wang C, Zhang M, Liu Y. Electrospun nanofibers of p-type NiO/n-type ZnO heterojunctions with enhanced photocatalytic activity. ACS Applied Materials and Interfaces. 2010;**2**:2915-2923. DOI: 10.1021/am100618h

[26] Chuangchote S, Jitputti J, Sagawa T, Yoshikawa S. Photocatalytic activity for hydrogen evolution of electrospun TiO_2 nanofibers. ACS Applied Materials and Interfaces. 2009;**1**:1140-1143. DOI: 10.1021/am9001474

[27] Kim H, Choi Y, Kanuka N, Kinoshita H, Nishiyama T, Usami T. Preparation of Pt-loaded TiO_2 nanofibers by electrospinning and their application for WGS reactions. Applied Catalysis A: General. 2009;**352**:265-270. DOI: 10.1016/j.apcata.2008.10.016

[28] Wang ZG, Wan LS, Liu ZM, Huang XJ, Xu ZK. Enzyme immobilization on electrospun polymer nanofibers: An overview. Journal of Molecular Catalysis B: Enzymatic. 2009;**56**:189-195. DOI: 10.1016/j.molcatb.2008.05.005

[29] Wan LS, Ke BB, Wu J, Xu ZK. Catalase immobilization on electrospun nanofibers: Effects of porphyrin pendants and carbon nanotubes. The Journal of Physical Chemistry C. 2007;**111**:14091-14097. DOI: 10.1021/jp070983n

[30] Barakat NA, El-Newehy M, Al-Deyab SS, Kim HY. Cobalt/copper-decorated carbon nanofibers as novel non-precious electrocatalyst for methanol electrooxidation. Nanoscale Research Letters. 2014;**9**:2. DOI: 10.1186/1556-276X-9-2

[31] Ye XR, Lin Y, Wang C, Engelhard MH, Wang Y, Wai CM. Supercritical fluid synthesis and characterization of catalytic metal nanoparticles on carbon nanotubes. Journal of Materials Chemistry. 2004;**14**(5):908-913. DOI: 10.1039/B308124A

[32] Yang W, Yang S, Guo J, Sun G, Xin Q. Comparison of CNF and XC-72 carbon supported palladium electrocatalysts for magnesium air fuel cell. Carbon. 2007;**45**:397-401. DOI: 10.1016/j.carbon.2006.09.003

[33] Guo DJ, Li HL. Electrochemical synthesis of Pd nanoparticles on functional MWNT surfaces. Electrochemistry Communications. 2004;**6**:999-1003. DOI: 10.1016/j.elecom.2004.07.014

[34] Li D, Xia Y. Electrospinning of nanofibers: Reinventing the wheel? Advanced Materials. 2004;**16**:1151-1170. DOI: 10.1002/adma.200400719

[35] Park JY, Lee IH, Bea GN. Optimization of the electrospinning conditions for preparation of nanofibers from polyvinylacetate (PVAc) in ethanol solvent. Journal of Industrial and Engineering Chemistry. 2008;**14**:707-713. DOI: 10.1016/j.jiec.2008.03.006

[36] Inagaki M, Yang Y, Kang F. Carbon nanofibers prepared via electrospinning. Advanced Materials. 2012;**24**:2547-2566. DOI: 10.1002/adma.201104940

[37] Kim C, Yang KS. Electrochemical properties of carbon nanofiber web as an electrode for supercapacitor prepared by electrospinning. Applied Physics Letters. 2003;**83**:1216-1218. DOI: 10.1063/1.1599963

[38] Kuzmenko V, Naboka O, Gatenholm P, Enoksson P. Ammonium chloride promoted synthesis of carbon nanofibers from electrospun cellulose acetate. Carbon. 2014;**67**:694-703. DOI: 10.1016/j.carbon.2013.10.061

[39] Kim C, Jeong YI, Ngoc BTN, Yang KS, Kojima M, Kim YA, et al. Synthesis and characterization of porous carbon nanofibers with hollow cores through the thermal treatment of electrospun copolymeric nanofiber webs. Small. 2007;**3**:91-95. DOI: 10.1002/smll.200600243

[40] Brandrup J, Immergut EH, Grulke EA, Abe A, Bloch DR, editors. Polymer Handbook. Vol. 89. New York: Wiley; 1999

[41] Wei K, Kim KO, Song KH, Kang CY, Lee JS, Gopiraman M, et al. Nitrogen- and oxygen-containing porous ultrafine carbon nanofiber: A highly flexible electrode material for supercapacitor. Journal of Materials Science and Technology. 2017;**33**:424-431. DOI: 10.1016/j.jmst.2016.03.014

[42] Chang WM, Wang CC, Chen CY. Fabrication of ultra-thin carbon nanofibers by centrifuged-electrospinning for application in high-rate supercapacitors. Electrochimica

Acta. 2019;**296**:268-275. DOI: 10.1016/j.electacta.2018.08.048

[43] Endo M, Kim YA, Ezaka M, Osada K, Yanagisawa T, Hayashi T, et al. Selective and efficient impregnation of metal nanoparticles on cup-stacked-type carbon nanofibers. Nano Letters. 2003;**3**:723-726. DOI: 10.1021/nl034136h

[44] Atchison JS, Zeiger M, Tolosa A, Funke LM, Jäckel N, Presser V. Electrospinning of ultrafine metal oxide/carbon and metal carbide/carbon nanocomposite fibers. RSC Advances. 2015;**5**:35683-35692. DOI: 10.1039/C5RA05409E

[45] Chen L, Hong S, Zhou X, Zhou Z, Hou H. Novel Pd-carrying composite carbon nanofibers based on polyacrylonitrile as a catalyst for Sonogashira coupling reaction. Catalysis Communications. 2008;**9**:2221-2225. DOI: 10.1016/j.catcom.2008.05.002

[46] Zhang P, Shao C, Zhang Z, Zhang M, Mu J, Guo Z, et al. In situ assembly of well-dispersed Ag nanoparticles (AgNPs) on electrospun carbon nanofibers (CNFs) for catalytic reduction of 4-nitrophenol. Nanoscale. 2011;**3**:357-3363. DOI: 10.1039/C1NR10405E

[47] Yu B, Liu Y, Jiang G, Liu D, Yu W, Chen H, et al. Preparation of electrospun Ag/g-C_3N_4 loaded composite carbon nanofibers for catalytic applications. Materials Research Express. 2017;**4**:015603

[48] Ghouri ZK, Barakat NA, Obaid M, Lee JH, Kim HY. Co/CeO_2-decorated carbon nanofibers as effective non-precious electro-catalyst for fuel cells application in alkaline medium. Ceramics International. 2015;**41**:2271-2278. DOI: 10.1016/j.ceramint.2014.10.031

[49] Formo E, Peng Z, Lee E, Lu X, Yang H, Xia Y. Direct oxidation of methanol on Pt nanostructures supported on electrospun nanofibers of anatase. The Journal of Physical Chemistry C. 2008;**112**:9970-9975. DOI: 10.1021/jp803763q

[50] Sonogashira K. Development of Pd–Cu catalyzed cross-coupling of terminal acetylenes with sp_2-carbon halides. Journal of Organometallic Chemistry. 2002;**653**:46-49. DOI: 10.1016/S0022-328X(02)01158-0

[51] Gopiraman M, Deng D, Saravanamoorthy S, Chung IM, Kim IS. Gold, silver and nickel nanoparticle anchored cellulose nanofiber composites as highly active catalysts for the rapid and selective reduction of nitrophenols in water. RSC Advances. 2018;**8**:3014-3023. DOI: 10.1039/C7RA10489H

[52] Chang YC, Chen DH. Catalytic reduction of 4-nitrophenol by magnetically recoverable Au nanocatalyst. Journal of Hazardous Materials. 2009;**165**:664-669. DOI: 10.1016/j.jhazmat.2008.10.034

[53] Gopiraman M, Saravanamoorthy S, Chung IM. Highly active human-hair-supported noble metal (Ag or Ru) nanocomposites for rapid and selective reduction of p-nitrophenol to p-aminophenol. Research on Chemical Intermediates. 2017;**43**:5601-5614. DOI: 10.1007/s11164-017-2950-3

[54] Gopiraman M, Deng D, Kim BS, Chung IM, Kim IS. Three-dimensional cheese-like carbon nanoarchitecture with tremendous surface area and pore construction derived from corn as superior electrode materials for supercapacitors. Applied Surface Science. 2017;**409**:52-59. DOI: 10.1016/j.apsusc.2017.02.209

Section 3

Electrospraying

Chapter 4

Effect of Spray-Drying and Electrospraying as Drying Techniques on Lysozyme Characterisation

Ijeoma Abraham, Eman Ali Elkordy, Rita Haj Ahmad, Zeeshan Ahmad and Amal Ali Elkordy

Abstract

The production of biopharmaceutical formulation incorporates several difficulties embracing their physical and chemical instabilities. In this study, two drying techniques, namely, spray-drying and electrospraying, were used to assess their application on lysozyme (as a model protein) without and with the use of betacyclodextrin. Samples were prepared in the ratio of 1:1 w/w (protein/betacyclodextrin), and several characterisation methods were applied to study the percentage (%) yield, morphology of the produced partials, thermal stability and biological activity of the protein. The results show the two drying methods led to different particle morphology as spherical-like shape was produced by spray-drying, while rodlike shape was generated by electrospraying with larger particle size. Lysozyme formulations produced by electrospraying were stable just directly after preparation, but after few weeks, those formulations showed visible aggregates. The biological activity of lysozyme was preserved by both drying techniques. In conclusion, both drying methods have different effects on the protein integrity and biological activity in which spray-drying shows more promising results.

Keywords: lysozyme, spray-drying, electrospraying, thermal stability, biological activity

1. Introduction

The majority of the FDA-approved protein-based medicines are delivered via conventional injection route (e.g. subcutaneous, intramuscular or intravenous). The pharmaceutical industries are faced with one of the most significant problems in protein manufacturing. One of the setbacks is the stability in the processing, manufacturing and storage of these therapeutic drugs. Solid dosage forms of therapeutic proteins could improve protein's bioavailability and stability during processing and storage. Various formulation techniques were applied utilising drying process aiming to develop a stable protein formulation (e.g. spray-drying, freeze drying, electrospraying, electrospinning, etc.). This chapter will be looking at the stability of lysozyme as a model of protein using spray-drying technique and electrohydrodynamic atomisation (EHDA) also known as electrospray. Very few publications have looked at the stability of lysozyme using EHDA technique. This challenge has necessitated the contribution towards this area.

1.1 Proteins

Proteins are macromolecules which require their native structure to be biologically active, and their conformation is very important in the development of protein drugs. They may denature with structural changes under stress, and there will be a loss of activities in the molecules. Examples of stresses are heat, elevated temperature, pressure, surface adsorption and pH [1]. Proteins undergo physical and chemical degradation; examples of physical degradation include aggregation, precipitation and unfolding as updated in Hui et al. [2] which involves the transition of protein from its native state to an unfolded state and will follow a significant loss in the function of a protein which generally will cause an unstable solution during the processing, manufacturing and storage.

1.2 Structure of proteins

Proteins consist of chains or small units of amino acid also known as amino acid polymers or building blocks [3] which contain the backbone or main chain of repeated units with attachments of variable side chains and are linked by peptide bonds. Each protein has a unique sequence of the side chains which determines the characteristics of the individual chain. There is a free carboxyl terminal at the end and a free amino terminal at the other end of every protein except for few cyclic polypeptides. The amino acid sequence is given in order of N-terminal to the C-terminal [4].

They are macromolecules heterogeneous in their native environment and are in most cases unstable. The ordering sequences in amino acids are referred to as the primary structure of protein and the secondary structure (α helix and β sheet); these are three-dimensional elements which all have an orientation of the protein backbone; tertiary structure is formed from secondary structural elements [5]; and quaternary structure comprises of several subunits with tertiary structures [6]. The configuration is determined by the native form followed by the assignment of α helix and β sheet which produces secondary conformation as these molecules are all linked by hydrogen bond [7].

This stability of the protein structure should include the three-dimensional state, the folded and the tertiary state which are all required for the biological activity. Although conformational stability is not enough rather, the protein must be able to find the folding pathway or its state within a short time from a denatured, unfolded conformation [8]. Folding maximises exposure of polar groups to the solvent and minimises exposure of non-polar groups. Protein unfolding is the transition from a native form to a denatured state [9]. When molecules are in aqueous solution, there is an equilibrium between folded which is the native state conformation and unfolded known as denatured. Native conformation stability is based on the relationship thermodynamically between ΔS and ΔH and the extent of ΔG-Gibbs free energy of the system. When the magnitude of ΔG is negative, it shows that there is a high stability of the native conformation than the denatured state which means that the greater the stability, the more negative is the ΔG. During protein unfolding, the main critical bonds required for protein stabilization are broken. Unfolding of proteins can take place at high temperatures, where the entropy is the main factor and the conformational entropy overpowers stabilizing forces. Hence, protein unfolding takes place. DSC takes measurement of ΔH of unfolding as a result of heat denaturation [10, 11].

1.3 Stability of proteins

Stability of protein is the achievement of a balance between stabilising and destabilising forces. The stabilising forces are caused by protein-solvent and intra-protein interaction, while the other is caused by substantial increase in entropy of unfolding [12].

One of the major challenges with the biopharmaceuticals is achieving protein stability.

There are three (3) types of stability:

1. Conformational stability

2. Chemical stability

3. Colloidal stability

Conformational stability describes the ability of the protein to maintain its native structure and be properly folded. Chemical stability is the stability among amino acid, covalent bond and different protein domains, while colloidal stability is the ability of the native structure of protein in a solution to avoid precipitation, aggregation and phase separation. However, it is necessary to maintain all types of stability during all the stages of development, processing and manufacturing.

The instability of protein necessitates the production in solid forms as solid proteins can be crystallised and dried [13] and its administration as a drug injection rather than through oral means like other drugs. Therefore, excipients also known as stabilisers are introduced to preserve the state and folding reversibility of proteins and reduce aggregation [14]. In effect to tackle protein degradation, necessary precautions and measures are taken to select the appropriate formulations for any excipients as well as the right technique for optimisation.

Proteins produced in different forms, like liquid formulations for medicines, are injectable due to the ease of preparation and are the most preferred, in manufacturing as well as the end users [15]. Unfortunately, it is not convenient due to the susceptibility of these proteins to the risk factors. This has instigated the consideration of the use of excipients as earlier mentioned in a dry solid state as important. Glass-forming agents such as saccharides, polyols and organic acids have been studied extensively over the years to stabilise spray-dried proteins in the solid state [16, 17]; these excipients stabilise the macromolecules by two primary mechanisms. Usually, the glass-forming ability of these excipients preserves the structure of proteins by trapping it in a rigid amorphous glass matrix [18, 19].

1.4 Lysozyme as a protein model

Lysozyme (1,4-β-N-acetylmuramidase) is the model of protein used in this study because lysozyme is one of the most potent proteins containing about 129 amino acids and has a high ionic strength. pH and the enzymes depend highly on their tertiary structure for maintaining their biological activity [20]. Lysozyme is folded into globular compact structure with a long cleft in the surface of the protein. The binding of the bacterial carbohydrate chain and the cleaving take place in the active site known as the cleft [21]; it has a molecular weight of 14.3 kDa and an isoelectric point of 10.7.

Lysozyme can be protected against bacteria through its activity as an enzyme as the role of an enzyme is to speed up the rate of chemical reaction in the body. Usually, polysaccharides are found in the cell walls of the affected bacteria which contain amine groups (NH_2) with sugar and side chains of sugar. Addition of water molecules to the sugar linkage causes lysozyme to degrade the polysaccharide, thereby causing it to break open. Its activity is a function of the ionic strength and pH. It is active between a pH of 6.0 and pH 9.0. Maximum activity is observed at a pH of 6.2 and ionic strength 0.02–0.100 M compared to 0.01–0.06 M at pH 9.2.

1.5 Drying techniques

1.5.1 Electrohydrodynamic atomisation

EHDA also known as electrospray is a drying technique used in the production of dry powder with the help of charges. It was first observed and recorded by Williams Gilbert in 1600 [22]. It is a process where a liquid imposes through a nozzle (which is connected to a voltage supply) at a certain flow rate, subjecting to high potential electric field. As a result, the meniscus is elongated to form a jet that breaks up into droplet under electrical force influence. Particles can then be collected at a collecting platform that is located ~20 cm under the spraying nozzle (**Figure 1**). Various spraying modes can be obtained depending on the strength of the electric stress and on the kinetic energy of the liquid leaving the nozzle for nanoparticle production. It is a well-practised technique for generating very fine droplets with monodispersed size from the liquid under the influence of electric forces [23]. It can achieve controlled monodispersity and morphology of particles without denaturation of bioactive molecules throughout the process, and this is possible with the use of emulsion. However, it has the potential to reduce or stop degradation of protein drugs and offer a strict control over particle morphology and size distribution. The principle of electrospray is based on the theory of charged droplets, 'stating that an electric field applied to a liquid droplet exiting a capillary is able to deform the interface of the droplet' [24]. Some literatures published state that electrospray is better than the other drying techniques because of its advantages as it does not require a high temperature, might use little or no emulsion also and may not need further drying. In a study [25], lysozyme as a model protein was encapsulated within biodegradable microparticles using coaxial electrospray and the authors concluded that electrospray could be a promising approach to encapsulate biomolecules [25]. Bock et al. have also done a review on electrospraying of polymer with therapeutic

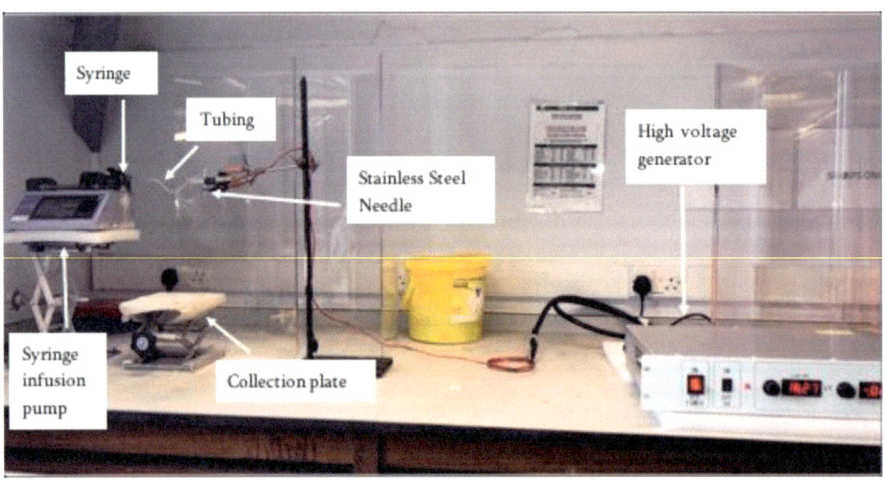

Figure 1.
Picture shows electrospray technique.

molecules as a state of the art and have concluded also that electrospraying technique may emerge as promising in the production of particles with entrapped therapeutic molecules that may be released as particle degradation occurs. So far, most of the research reports for electrospraying technique using proteins were based on using polymeric systems in order to encapsulate the protein either in protein carriers or nano- or microcapsules. For example, Suksamran et al. [26] successfully synthesised alginate microparticles (0.9–4 μm in diameter) with ~50% entrapment efficiency of bovine serum albumin that raised promises for oral drug delivery of proteins. Moreover, tristearin nanoparticles (~100–300 nm in diameter) were developed to aid the delivery of angiotensin II alongside clear core-shell particles with ~92 ± 1.8% encapsulation efficiency [27]. Electrosprayed core-shell microspheres (2–8 μm) with encapsulated bovine serum albumin as the core and an amphiphilic biodegradable polymer (poly(ε-caprolactone)-poly(aminoethyl ethylene phosphate) block copolymer) as the shell for protein delivery were generated by a single-step electrospraying. The protein release profiles of the microspheres exhibited steady release kinetics for a period of 3 weeks without a significant initial burst [28].

1.5.2 Spray-drying

Spray-drying is an established technology for the production of dried products from the liquid state. This method has gained increased interest in dry power formulation over the past decade, due to its potential simplicity, adaptability, scalability and cost-effectiveness [29, 30]. It is a method that has been studied in dry powder protein production involving the use of high temperature during the drying process. The principle of spray-drying method relies on atomisation of a drug solution that is pumped into a dry hot chamber in the form of droplets. By the influence of the heat, these droplets evaporate leading to dry solids in either powder, granules or agglomerated particles based on the chemical and physical properties of the feed in addition to the design of the used spray-dryer (**Figure 2**). One of the challenges of producing protein formulation utilising spray-drying method is the ability to destabilise the protein due to hot temperature and pressure during the production process compared to electrospray which uses charges for drying.

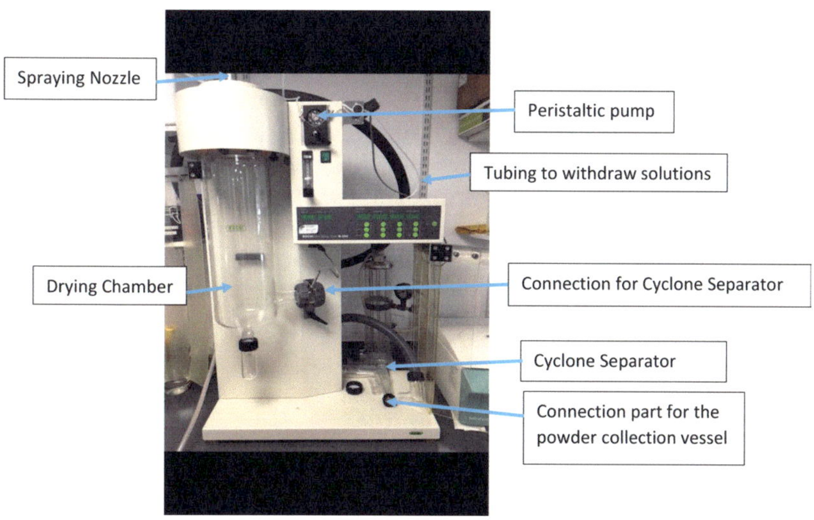

Figure 2.
Picture shows the spray-drying technique.

Various reports documented the use of spray-drying that produced protein formulations with high protein activity. For instance, the fabrication of uniform trehalose microparticles immobilised with trypsin showed ~97.7 ± 2.6% biological activity using an optimal trypsin/trehalose mass ratio of 1:1 [31]. Insulin microparticles were also designed by spray-drying method to bypass deposition in the extrathoracic region (mouth-throat). The total lung dose of >95% was achieved indicating a high degree of lung targeting [32]. Newly developed microencapsulated solid lipid nanoparticles containing papain (as a model protein) were prepared by spray-drying method. Papain was adsorbed onto glyceryl dibehenate and glyceryl tristearate solid lipid nanoparticle. The protein was found to be released from the particles to a large extent. Moreover, protein stability was reserved throughout spray-dried microsphere production [33].

This chapter aims to describe the effect of spray-dried and electrospray-dried formulations on lysozyme with and without excipients using various characterisation techniques like differential scanning calorimetry (DSC) to determine the thermal stability with lysozyme solid samples, UV-Vis spectroscopy for the enzymatic activity, dynamic light scattering for particle size, scanning electron microscope (SEM) for morphology of the particles and high-sensitivity DSC for thermal stability using lysozyme solutions and two drying (electrospray and spray-drying) techniques, to dry powders for characterisation.

2. Materials and methods

2.1 Materials

Chicken egg white lysozyme in lyophilised powder (90%), ethanol, *Micrococcus lysodeikticus* and betacyclodextrin were purchased from Sigma-Aldrich Chemicals Company. Sodium phosphate monobasic anhydrous, disodium hydrogen orthophosphate anhydrous and sodium chloride were obtained from Fisher Scientific Company.

2.2 Methods

2.2.1 Preparation of spray-dried protein

Aqueous lysozyme solutions (1% w/v) were spray-dried without and with an excipient (betacyclodextrin) via a BÜCHI Mini Spray Dryer B-290. Excipients were used at the concentrations of 1% w/v. The protein solution run through a silicone tubing of inner diameter of 4 mm, peristaltic feed pump (35%) to an 0.5 mm diameter atomising nozzle at rate of 20 ml/min and compressed air at rate of 600 l/h. In a glass chamber, protein solutions were sprayed at an inlet temperature of 130 ± 3°C, and outlet temperature was 73 ± 4°C. To minimise the effect of heat stress on the protein, a cooling water was distributed through a jacket around the nozzle. Spray-dried protein powder were collected by a high-performance cyclone separator and stored in glass vials at 3–4°C.

2.2.2 Preparation of electrosprayed lysozyme

Lysozyme solutions (1% w/v) were prepared using (80/20 v/v) ethanol/water. Electrosprayed formulations were prepared without and with excipients (1% w/w of betacyclodextrin). A syringe containing 5 ml of the protein solution was attached to a syringe infusion pump (Harvard Apparatus, Pump 11 Elite, USA) attached to a high-power voltage supply (Glassman High Voltage Supply, UK). The protein solution was run through a silicone tubing attached to a stainless steel needle (inner diameter 0.3 cm) at a flow rates of 15 μL/min. Atomised particles were collected on microscope slides which fitted 20 cm below the tip of the nozzle. A high voltage was used to maintain the voltage range of 9–18 kV. The electrospraying process was conducted at ambient temperature (23°C).

2.2.3 Characterisation of spray-dried and electrosprayed lysozyme

2.2.3.1 Scanning electron microscopy

Microscopic examination of spray-dried and electrosprayed formulations was investigated utilising scanning electron microscope (SEM) (Hitachi S3000-N variable pressure scanning electron microscope, Japan). Around 2–3 mg of dried lysozyme formulations were applied to a double-sided carbon tape (Agar Scientific, Stansted, UK), fixed on an aluminium stub. The dried powder was sputter coated with a mixture of gold/palladium using a Quorum Technology (Polaron Range) SC760 by subjecting powder to an argon atmosphere at about 10^{-1} mbar.

2.2.3.2 Differential scanning calorimetry (DSC)

The thermal stability of lysozyme, before and after processing, was evaluated in the solid form using DSC Q1000M TA instrument, England. Pure indium standard was utilised to calibrate the DSC instrument. Unprocessed, spray-dried and electrosprayed formulation (2–3 mg) was sealed in a DSC aluminium pan with lids and loaded under nitrogen at a flow rate of 50 ml/min. An empty sealed pan was used as a reference. Pans were scanned in the range of 0–300°C at a rate of 10.0°C/min.

2.2.3.3 High-sensitivity differential scanning calorimetry (HSDSC)

High-sensitivity DSC also known as VP-DSC was used in this study to determine the thermal behaviour and folding reversibility of lysozyme. The fresh sample 5 mg of each formulation (1 ml 0.1 M phosphate buffer at a pH of 6.24) and reference (0.1 M sodium phosphate buffer at a PH of 6.24) was degassed and injected into the cells using a syringe. The sample and reference were complete for maintenance of equal volumes, and the same amount of lysozyme was used in each run. The sample and reference were heated between 20 and 90°C at 1°C/min under pressure. The folding reversibility each, of the denatured protein, was evaluated by cycling temperature by carrying out three scans (up scan, 20–90°C; down scan, 90–20°C; and another up scan, 20–90°C). These engaged two or more scans at different temperatures. Furthermore, before the measurement of each sample, a baseline was run by loading both cells (sample and reference cells) with the reference; the baseline was later subtracted from the protein thermal data using the MicroCal VP-DSC, USA. The sample was analysed with Origin DSC analysis software by normalising the concentration and excesses in heat capacity. The calorimetric enthalpy changes which is the area under the peak (ΔH) and midpoint of the transition peak (T_m) were calculated.

2.2.3.4 Biological activity analysis of lysozyme

Utilising M501 Single Beam Scanning UV/Visible Spectrophotometer Camspec (Biochrom, UK), the enzymatic activity of lysozyme was investigated by determining the hydrolysis rate of β-1,4-glycosidic linkages between N-acetylglucosamine and N-acetylmuramic acid in bacterial cell walls. Following a procedure described by Haj Ahmad et al. [34], the activity of lysozyme was assessed. A 100 ml of *Micrococcus lysodeikticus* suspension was prepared by adding 20 mg of the bacteria added to 10 ml of 1% sodium chloride solution and 90 ml of potassium phosphate buffer 0.067 M, pH 6.6. The biological reaction of lysozyme was initiated by adding 0.5 ml of protein solution (1.5 μg/ml), prepared using the same buffer, to 5 ml of the bacterial suspension. The unit activity of the protein was monitored by measuring the reduction rate in the absorbance at 450 nm. The biological activity of lysozyme was assessed using the equation suggested by Shugar [35]:

$$\text{Activity (units/mg)} = \Delta 450_{\text{nm/min}}/0.001 \times \text{mg enzyme present in the mixture.} \quad (1)$$

2.2.3.5 Particle size distribution analysis

The particle size distribution of the dried particles was determined by dynamic light scattering technique using the Malvern Zetasizer (Nano ZSP, Malvern Instruments, UK). The particles were dispersed in ethanol containing 0.1% Tween® 80, which was selected to achieve suitable deagglomeration. The average particle size was measured at a scattering angle of 90° in three replicates for each sample.

3. Results and discussion

3.1 Determination of percentage yields for spray-dried and electrosprayed lysozyme

Table 1 shows the percentage yield of the spray-dried and electrosprayed lysozyme formulation collected. For spray-dried lysozyme, 29% of the preparation was achieved. The percentage yield was higher (~53%) for spray-dried lysozyme with betacyclodextrin compared to the spray-dried lysozyme in the absence of excipient. The addition of betacyclodextrin reduced the deposition of spray-dried lysozyme on the wall of the spray-drying chamber and cyclone separator. With no significant differences, lysozyme formulation produced with electrospraying produced ~30%. However, a major reduction was noticed for electrospraying lysozyme with betacyclodextrin (~25%). Around 30–40% of product yield is typically expected for spray-drying formulations utilising benchtop spraying system [36]. Increasing the percentage yield can be achieved by introducing polymeric excipients with high glass transition temperature in addition to optimise the drying condition used during the spray-drying [37]. A large-scale production is feasible in pharmaceutical industries for spray-drying formulations by using a large scalable spray-drier that would generate the highest possible yield [38]. This is also feasible for electrospraying by using, for example, a multi-tip emitter to improve the potential upscale electrospraying [39]. For electrospraying technique, the process parameters, such as feeding flow rate, voltage supply and the distance between the tip of the nozzle and the collecting platform, would affect the physical properties of the produced particles. It's worth mentioning that the time consumed by each process to generate the quantity outlined in **Table 1** was significantly different. Spray-drying process required 15–20 min for the whole sample, while more than 6 h was consumed to finish with all of the electrosprayed sample.

3.2 Microscopic examination of spray-dried and electrosprayed protein particles

Figure 3 and **Table 1** show the SEM images and particle size of spray-dried and electrosprayed formulations, respectively. The process of protein preparation

Formulation	Percentage (%) yield	Biological activity (%)	Tm (°C)[a]	Apparent Tm (°C)[b]	Mean diameter (μm) (mean ± SD)
Unprocessed lysozyme		100	74.32 ± 0.16	199.06	1.4 ± 0.04
SD lysozyme	29	90.1	72.46 ± 0.38	201.30	2.3+ 0.06
ESR lysozyme	30	113	74.44 ± 0.22	222.18	1.6 ± 0.19
SD (1:1 w/w) lysozyme with betacyclodextrin	53	87.06	74.36 ± 0.26	No transition detected	2.7 ± 0.5
ESR (1:1 w/w) lysozyme with betacyclodextrin	25	100	74.51 ± 0.52	223.26	3.0 ± 0.91

[a]*For aqueous protein formulations.*
[b]*For solid protein formulations.*
SD, spray-dried; ESR, electrosprayed

Table 1.
Shows percentage (%) yield, biological activity, denaturation temperature, Tm and particle size of dried lysozyme particles.

had a significant impact on the morphology and size of the spray-dried and electrosprayed lysozyme particles. As a result, it will impact the choice of the protein delivery system and route of drug administration. Protein particles produced by spray-drying technique were smooth and spherical in shape (**Figure 3A**) with a size of 2.3 + 0.06 μm, while mainly rodlike shape particles with few spherical particles were produced via electrospraying process (1.6 ± 0.19 μm). The inclusion of betacyclodextrin had no significant effect on the shape of electrosprayed particles (3.0 ± 0.91 μm), while solid dimple spheres were observed for spray-dried lysozyme with betacyclodextrin formulation (2.7 ± 0.5 μm) (**Figure 3**). For spray-drying process, Prinn et al. [40] suggested four categories to classify the morphology of the spray-dried particles: (I) smooth spheres, (II) dimpled or collapsed particles, (III) raisin like particles and (IV) highly crumpled folded particles. Accordingly, spray-dried lysozyme without betacyclodextrin would fall into class I, and spray-dried lysozyme with betacyclodextrin would fall into class II. The rate of drying has a crucial impact on the morphology of the spray-dried particles; faster drying would result in more dimpled particles [41]. However, the spherical shape of the particles will not guaranty a 100% biological activity as will be discussed below. Moreover, it has been recognised that the optimal aerodynamic particle size distribution of particles intended for pulmonary delivery should be within the range of 1–5 μm [42]. Accordingly, all particles produced in this research would fit within this range and thus are suitable for pulmonary delivery.

The electrospraying configuration used in this study consists of one conducting needle, and one voltage was applied. A stable Taylor cone usually forms at the tip of the needle generating nearly uniform particles with distinct morphology. The incorporated protein disperses equally with the solution (without and with the excipient) improving the amorphous nature and bioavailability (Mehta et al. [22]). Flow rate is highly related to the particle size of the electrosprayed particles. It documented that larger flow rates tend to generate smaller particle size [43]. The flow rate used in this study considered as low flow rate that might explain the large particle size of electrosprayed lysozyme especially with betacyclodextrin (**Table 1**, **Figure 3**). Also, a study conducted by Gomez et al. [44], electrospraying of insulin, suggested reducing protein concentration in the

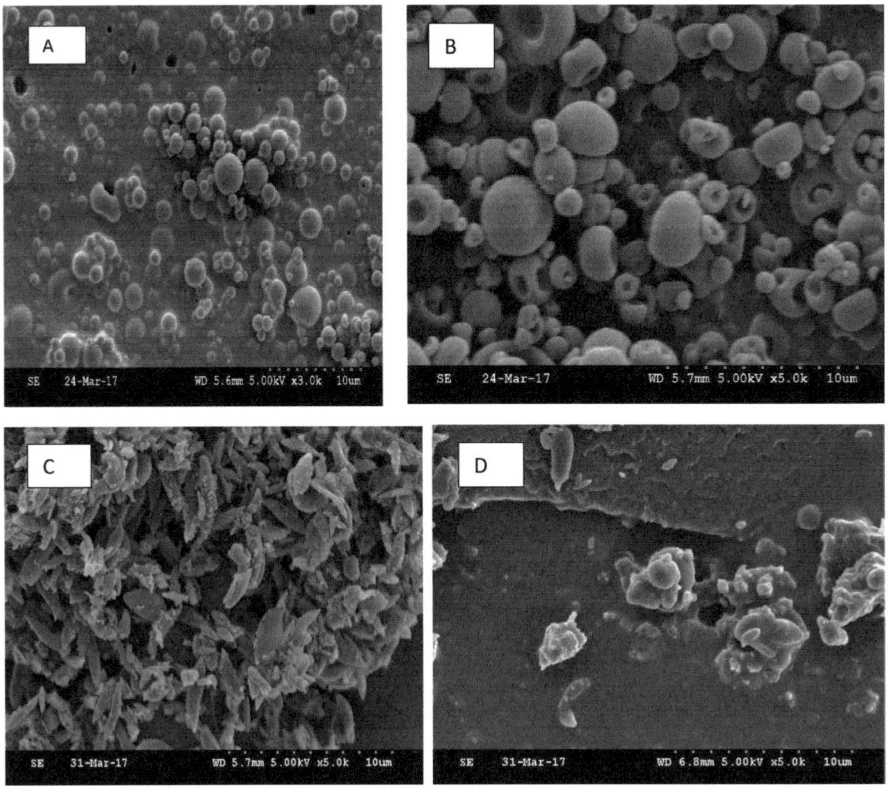

Figure 3.
SEM images for (A) spray-dried lysozyme, (B) spray-dried lysozyme with betacyclodextrin, (C) electrosprayed lysozyme and (D) electrosprayed lysozyme with betacyclodextrin.

solvent or the flow rate in order to achieve a smaller particle size. A future study with various lysozyme concentrations and higher and lower flow rates will be conducted. The particle morphology of protein formulations and size play a crucial role in the aerodynamic properties and performance of aerosol applications. Smooth spherical particles are more preferred than other particle shapes as they might result in much lower aerodynamic particle diameter in comparison with dense particles.

3.3 Differential scanning calorimetry (DSC)

DSC determines the variation in heat flow between the protein sample and an empty sealed pan that was used as a reference cell. Throughout a thermal event in the protein preparation, the operation of the system will conduct heat to, or from, the protein sample pan to maintain an equal temperature in both sample and reference pans. Thermal profiles of unprocessed, spray-dried and electrosprayed lysozyme formulations without excipients are presented in **Figure 4**. This represents the heat flow as a function of temperature and illustrates the apparent denaturation temperature T_m values of all protein preparations (**Table 1**). In the DSC thermogram scans, two endothermic peaks can be observed. The first endothermic broad peak (~120°C) indicates that the preparations contain some amount of water [45, 46]. The second endothermic peak around ~199°C represents the apparent denaturation transition of the protein in which the peak maximum was considered the apparent denaturation temperature [21].

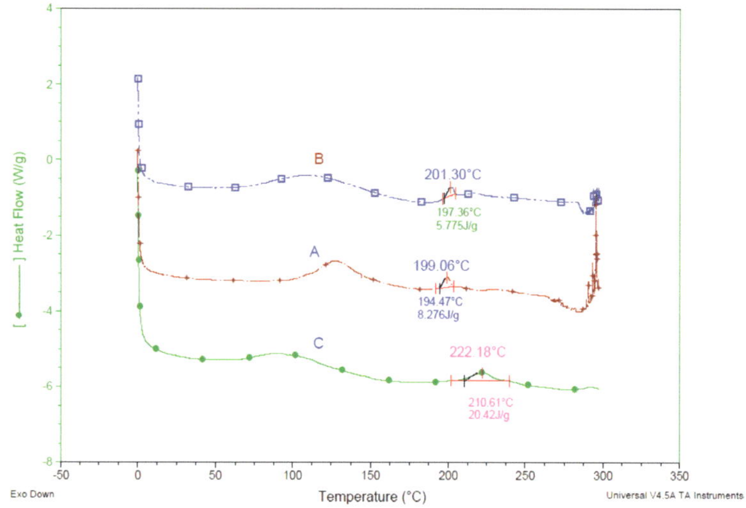

Figure 4.
DSC thermograms for unprocessed, spray-dried and electrosprayed lysozyme.

Spray-drying of lysozyme has improved the denaturation temperature of lysozyme by about 2°C as compared with unprocessed lysozyme (**Table 1**). However, this transition was not significant. The presence of betacyclodextrin within spray-dried sample indicated an amorphous protein content as no peak was detected around 199°C; this was in contrast with other publications which showed transition for spray-dried lysozyme with betacyclodextrin. Around 22°C increase of the denaturation temperature of lysozyme was observed for electrosprayed protein without any excipient (**Table 1**, **Figure 4**). Moreover, electrospraying of lysozyme with betacyclodextrin led to an increase (by ~23°C) of the apparent protein denaturation temperature compared to the unprocessed lysozyme. The results for electrospraying formulations suggest an increase of lysozyme thermal stability as they exhibited increase in denaturation temperatures indicating the effect of the process and the excipient on the protein's integrity, and this is in consistent with the biological activity results for freshly dried protein samples.

HSDSC offers information about protein folding and stability by measuring the thermodynamic parameters in solution forms which have an impact on protein folding-unfolding transitions [47]. Subsequently, this method was used to assess the thermal stability of the prepared protein solutions after processing. **Table 1** displays the HSDSC results for denaturation temperature of unprocessed, spray-dried and electrosprayed lysozyme preparations. The results (**Table 1**) show that the transition temperature (T_m) of the spray-dried formulation in the absence of betacyclodextrin reveals lower thermal stability (T_m was ~72°C) than the unprocessed lysozyme and other formulations (T_m was ~74°C). Electrosprayed samples have maintained the thermal stability of the protein after processing (T_m was ~74°C). However, after 6 months of storage of electrosprayed samples in solid forms, it was noticed that those samples showed visible aggregates upon addition of aqueous solutions, while spray-dried lysozyme samples were soluble in the aqueous solutions.

3.4 Biological activity of lysozyme

The most crucial parameter in any protein formulation is to maintain the biological activity of the protein. This will give an indication about the integrity, foldability and stability of the protein. Biological activity test was run for lysozyme before and after processing. Solid protein samples were reconstituted, and the biological activity was expressed as a percentage that is relative to the control protein (activity of unprocessed protein was 100%). **Table 1** displays the biological activity results for spray-dried and electrosprayed lysozyme formulations. Spray-drying of lysozyme without and with betacyclodextrin led to about 10% reduction of lysozyme activity as compared with the unprocessed lysozyme. On the other hand, electrospraying of lysozyme better maintained protein biological activity compared to spray-dried protein formulations (**Table 1**). However, aggregate formation in solutions was noticed after 6 months of storage of electrosprayed lysozyme solid samples, meaning that proper storage conditions require to be maintained for long-term stability of dried lysozyme samples prepared by electrospraying technique.

The outcomes of this study could be explained on the basis that spray-drying process has perturbed the tertiary structure of lysozyme, thus reducing the biological activity of the protein; these results are in agreement with Haj Ahmad et al. [21] and Hulse et al. [45]. Lu et al. [48] reported that the type and percentage of the excipient used would influence the stability of protein formulation. The addition of betacyclodextrin to the spray-drying formulation did not have any improvement as ~87% biological activity of lysozyme was achieved. Sustaining the tertiary structure of lysozyme is critical for its full activity. One of the vital factors for this is the level of hydration at the active site cleft of the protein. Nagendra et al. [49] suggested that the active site in the lysozyme is heavily hydrated and ~0.2 grammes of water/g protein is required for the protein to maintain its biological activity with at least 9.4% of moisture content. Otherwise, the active site cleft will shrink, and thus the protein will inactivate [49]. On the other hand, 100% of the protein biological activity was observed for freshly electrosprayed lysozyme without and with betacyclodextrin suggesting the full hydration of the active site cleft in the protein. A study conducted by Gomez et al. [44] using insulin as a model protein reported electrospraying technique was gentle not to hinder the biological activity of insulin. The overall results suggest that different ways of drying (e.g. heating like in spray-drying and electric charge like in electrospraying) have different influences on the protein biological activity. The structure of proteins embraces several weak interactions (hydrogen bonging and electrostatic). These interactions can easily be affected by subjecting to physical and chemical stimuli that would include various processing conditions of conventional processing techniques (e.g. spray-drying) where high temperature or stress is applied. High temperature tends to denature sensitive biomolecules, such as proteins. Atomisation of materials by electrospraying relies mainly on evaporation of the solvent to generate dry particles without affecting the characteristics of proteins such as biological activity. While denaturation by the effect of temperature can be avoided, some denaturation and degradation might be initiated by the shear stress in the nozzle tip through which the aqueous solvent is infused [22, 37, 50, 51]. In this study, the biological activity of freshly dried lysozyme forms was conserved by using electrospraying method compared with spray-drying technique. However, aggregate formation in solutions was noticed after 6 months of storage of electrosprayed lysozyme solid samples, meaning that proper storage conditions require to be maintained for long-term stability of dried lysozyme samples prepared by electrospraying technique. Accordingly, there is a need for further study of the stability of the protein particles at various temperature and humidity conditions.

4. Conclusion

It was observed that electrospray lysozyme formulations without and with the used excipient looked promising compared to unprocessed and spray-dried formulations. However, it was observed that proper storage for electrosprayed samples is needed as the protein degraded after some time of storage. Spray-dry and electrospray seem to have some positive and negative effects on the lysozyme depending on the technique and its used parameters.

Acknowledgements

This study was partially supported by EPSRC EHDA network for part funding and the ability to use the EHDA equipment for this research.

Author details

Ijeoma Abraham[1], Eman Ali Elkordy[2], Rita Haj Ahmad[3], Zeeshan Ahmad[3] and Amal Ali Elkordy[1*]

1 School of Pharmacy and Pharmaceutical Science, University of Sunderland, UK

2 Faculty of Medicine, Imam Mohammad Ibn Saud Islamic University, Saudi Arabia

3 Leicester School of Pharmacy, De Montfort University, Leicester, UK

*Address all correspondence to: amal.elkordy@sunderland.ac.uk

References

[1] Ohtake S, Yoshiko K, Arakawa T. Interactions of formulation excipients with proteins in solution and in dried state. Advanced Drug Delivery Reviews. 2011;**63**(13):1053-1073

[2] Lim H-P, Tey B-T, Chan E-S. Particle designs for the stabilisation and controlled-delivery of protein drugs by biopolymers: A case study on insulin. Journal of Controlled Release. 2014;**186**:11-21

[3] George CL. What is Protein? 1992. Available from: https://digitalcommons.usu.edu/cgi/viewcontent.cgi?article=1612&context=extension_histall [Accessed: March 6, 2019]

[4] Lesk MA. Introduction to Protein Science Architecture Function and Genomics. New York: Oxford University Press; 2004

[5] Cleland JL, Craik SC. Protein Engineering Principles and Practice. Canada: John Westley & Sons Publication; 1996. pp. 4-5

[6] Saito N, Kobayashi Y. The Physical Foundation of Protein Architecture. International Journal of Modern Physics B. 1999;**13**:2431-2529

[7] Elkordy AA, Forbes RT, Barry BW. Study of protein conformational stability and integrity using calorimetry and FT-Raman spectroscopy correlated with enzymatic activity. European Journal of Pharmaceutical Sciences. 2008;**33**(2):177-190

[8] Elkordy AA, Forbes RT, Barry BW. Integrity of crystalline lysozyme exceeds that of a spray-dried form. International Journal of Pharmaceutics. 2002;**247**:79-90

[9] Kocherbitov V, Arnebrant T. Hydration of thermally denatured lysozyme studied by sorption calorimetry and differential scanning calorimetry. The Journal of Physical Chemistry. 2006;**110**:10144-10150

[10] Gill P, Moghadam TT, Ranjbar B. Differential scanning calorimetry technique: Application in biology and nanoscience. Journal of Biomolecular Techniques. 2010;**21**:167-193

[11] Privalov PL, Dragan AI. Microcalorimetry of biological macromolecules. Journal of Biophysical Chemistry. 2007;**126**:16-24

[12] Challener CA. Excipient selection for protein stabilization. The Journal of Pharmacy Technology. 2015;**3**:35-39

[13] Forbes RT, Barry BW, Elkordy AA. Preparation and characterisation of spray dried and crystallised trypsin: FT-Raman study to detect protein denaturation after thermal stresses. European Journal of Pharmaceutical Sciences. 2007;**30**:315-323

[14] Moggridge J, Biggar K, Dawson N, Storey KB. Sensitive detection of immunoglobulin G stability using in real-time isothermal differential scanning Fluorimetry: Determinants of protein stability for antibody-based therapeutics. Technology in Cancer Research & Treatment. 2017;**16**(6):997-1005. DOI: 10.1177/1533034617714149

[15] Siew A. Freeze-drying protein formulation. The Journal of Pharmacy Technology. 2014;**38**:5

[16] Amorij JP, Huckriede A, Wilschut J, Frijlink H, Hinrichs W. Development of stable influenza vaccine powder formulations: Challenges and possibilities. Pharmaceutical Research. 2008;**25**:1256-1273

[17] Chang LL, Pikal MJ. Mechanisms of protein stabilisation in the solid state. Journal of Pharmaceutical Sciences. 2009;**98**:2886-2908

[18] Conrad PB, Miller DP, Cielenski PR, de Pablo JJ. Stabilization and preservation of lactobacillus acidophilus in saccharide matrices. Cryobiology. 2000;**41**:17-24

[19] Vehring R. Pharmaceutical particle engineering via spray drying. Pharmaceutical Research. 2008;**25**:999-1022

[20] Ghaderi R, Carlfors J. Biological activity of lysozyme after entrapment in poly (d, l-lactide-co-glycolide)-microspheres. Pharmaceutical Research Journal. 1997;**14**(11):1556-1562

[21] Haj Ahmad R, Elkordy AA, Chaw CS, Moore A. Compare and contrast the effects of surfactants (Pluronic®F-127 and Cremophor®EL) and sugars (β-cyclodextrin and inulin) on properties of spray dried and crystallised lysozyme. European Journal of Pharmaceutical Sciences. 2013;**49**(4):519-534

[22] Mehta P, Zaman A, Smith A, Rasekh M, Haj Ahmad R, Arshad MS, et al. Broad scale and structure fabrication of healthcare materials for drug and emerging therapies via electrohydrodynamic techniques. Advances in Therapy. 2018;**1800024**:1-27

[23] Xie J, Jiang J, Davoodi P, Srinivasan MP, Wang CH. Electrohydrodynamic atomisation: A two-decade effort to produce and process micro−/nano particulate materials. Chemical Engineering Science. 2015;**125**:32-57

[24] Bock N, Dargaville TR, Woodruff MA. Electro spraying of polymers with therapeutic molecules: State of the art. Progress in Polymer Science. 2012;**37**:1510-1551

[25] Xie J, Ng WJ, Lee LY, Wang CH. Encapsulation of protein drugs in biodegradable microparticles by co-axial electrospray. Journal of Colloid and Interface Science. 2008;**317**:469-476

[26] Suksamran T, Opanasopit P, Rojanarata T, Ngawhirunpat T, Ruktanonchai U, Supaphol P. Biodegradable alginate microparticles developed by electrohydrodynamic spraying techniques for oral delivery of protein. Journal of Microencapsulation. 2009;**26**(7):563-570

[27] Rasekh M, Young C, Roldo M, Lancien F, Le Mével J-C, Hafizi S, et al. Hollow-layered nanoparticles for therapeutic delivery of peptide prepared using electrospraying. Journal of Materials Science. Materials in Medicine. 2015;**26**:256

[28] Wu Y, Liao I-C, Kennedy SJ, Du J, Wang J, Leong KW, et al. Electrosprayed core-shell microspheres for protein delivery. Chemical Communications. 2010;**46**:4743-4745

[29] Fourie P, Germishuizen W, Wong Y-L, Edwards D. Spray drying TB vaccines for pulmonary administration. Expert Opinion on Biological Therapy. 2008;**8**:857-863

[30] Lee YY, Wu JX, Yang M, Young PM, van den Berg F, Rantanen J. Particle size dependence of polymorphism in spray-dried mannitol. European Journal of Pharmaceutical Sciences. 2011;**44**:41-48

[31] Zhang S, Lei H, Gao X, Xiong X, Wu WD, Wu Z, et al. Fabrication of uniform enzyme-immobilized carbohydrate microparticles with high enzymatic activity and stability via spray drying and spray freeze drying. Powder Technology. 2018;**330**:40-49

[32] Ung KT, Rao N, Weers JG, Huang D, Chan HK. Design of spray dried insulin microparticles to bypass deposition in the extrathoracic region and maximize total lung dose. International Journal of Pharmaceutics. 2016;**511**(2):1070-1079

[33] Gasper DP, Serra C, Lino PR, Goncalves L, Taboada P, Remunan-Lopez C, et al. Microencapsulated SLN:

An innovative strategy for pulmonary protein delivery. International Journal of Pharmaceutics. 2017;**516**(1-2):231-246

[34] Haj Ahmad R, Chen YT, Elkordy A. An overview for the effects of lactitol, gelucire 44/14 and copovidone on lysozyme biological activity. European Journal of Biomedical and Pharmaceutical Sciences. 2015;**2**(3):1328-1339

[35] Shugar D. Measurement of lysozyme activity and ultraviolet inactivation of lysozyme. Biochimica et Biophysica Acta. 1952;**8**:302

[36] Hulse WL, Forbes RT, Bonner MC, Getrost M. Influence of protein on mannitol polymorphic form produced during co-spray drying. International Journal of Pharmaceutics. 2009;**382**(1-2):67-72

[37] Ziaee A, Albadarin AB, Padrela L, Femmer T, O'Reilly E, Walker G. Spray drying of pharmaceuticals and biopharmaceuticals: Critical parameters and experimental process optimization approaches. European Journal of Pharmaceutical Sciences. 2019;**127**:300-318

[38] Haj Ahmad R, Mamayusupov M, Elkordy EA, Elkordy AA. Influences of copolymers (Copovidone, Eudragit® RL PO and Kollicoat® MAE 30 DP) on stability and bioactivity of spray-dried and freeze-dried lysozyme. Drug Development and Industrial Pharmacy. 2016;**42**(12):2086-2096

[39] Haj-Ahmad R, Rasekh M, Nazari K, Onaiwu EV, Yousef B, Morgan S, et al. Stable increased formulation atomization using a multi-tip nozzle device. Drug Delivery and Translational Research. 2018;**8**(6):1815-1827

[40] Prinn KB, Costantino HY, Tracy M. Statistical modelling of protein spray drying at the lab scale. AAPS PharmSciTech. 2002;**3**(1):E4

[41] Wang FJ, Wang CH. Effects of fabrication conditions on the characteristics of etanidazol spray-dried microspheres. Journal of Microencapsulation. 2002;**19**(4):495-510

[42] Hickey AJ. Pharmaceutical Inhalation Aerosol Technology. 2nd ed. In: Applications and Advances. CRC Press; 2016

[43] Almeria B, Deng W, Fahmy TM, Gomez A. Controlling the morphology of electrospray-generated PLGA microparticles for drug delivery. Journal of Colloid and Interface Science. 2010;**343**(1):125-133

[44] Gomez A, Bingham D, De Juan L, Tang K. Production of Protein Nanoparticles by Electrospraying; Journal of Aerosol Science. 1998;**29**:561-574

[45] Hulse LW, Forbes TR, Bonner CM, Getrost M. Do co-spray dried excipients offer better lysozyme stabilization than single excipients? European Journal of Pharmaceutical Sciences. 2008;**33**:294-305

[46] Kayaci F, Uyar T. Electrospun zein nanofibers in incorporating cyclodextrins. Carbohydrate Polymers. 2012;**90**:558-568

[47] Cooper A, Johnson CM. Chapter 10: Differential scanning calorimetry. In: Jones C, Mulloy B, Thomas AH, editors. Methods in Molecular Biology. Vol. 22: Microscopy, Optical Spectroscopy, and Macroscopic Techniques. Totowa, N.J: Humana Press; 1994. pp. 125-136

[48] Lu J, Wang X-J, Liu Y-X, Ching C-B. Thermal and FTIR investigation of freeze-dried protein-excipient mixtures. Journal of Thermal Analysis and Calorimetry. 2007;**89**(3):913-919

[49] Nagendra HG, sukumar N, Vijayan M. Role of water in plasticity,

stability, and action of proteins: The crystal structures of lysozyme at very low levels of hydration. Proteins Structure Function and Bioinformatics. 1998;**32**:229-240

[50] Sridhar R, Lakshminarayanan R, Madhaiyan K, Barathi VA, Limh KHC, Ramakrishna S. Electrosprayed nanoparticles and electrospun nanofibers based on natural materials: Applications in tissue regeneration, drug delivery and pharmaceuticals. Chemical Society Reviews. 2015;(44):790

[51] Yurteri CU, Hartman RPA, Marijnissen JCM. Producing pharmaceutical particles via electrospraying with an emphasis on nano and nano structured particles—A review. Kona Powder and Particle Journal. 2010;**28**:91-115

Chapter 5

Using the iESP Installed on the Space Station Moving in an Irregular Gravitational Field of the Asteroids Eros and Gaspra

Olga Starinova, Andrey Shornikov and Elizaveta Nikolaeva

Abstract

The gravitational field of asteroids and comets often has a complex dynamically changing shape. The study of the gravitational field of such bodies is essential for the design of missions, especially for missions involving maneuvering in close proximity, landing, and takeoff from the surface of a celestial body. This chapter deals to the problem of spacecraft's controlled motion near objects with irregular gravitational fields. We propose to use a propulsion system based on ion electrospray propulsions (iESP), which usually used to control the spacecraft's motion relative to the mass center. The chapter describes the gravitational field of an asteroid as the superposition of the fields of two rotating point masses. The proposed control laws made it possible to stabilize the orbit near the asteroid surface. We performed the simulation of motion close to two relatively well-studied asteroids Eros and Gaspra. The results of the simulation demonstrated that the using of iESP as the auxiliary propulsion systems is sufficient to control the spacecraft's motion in irregular gravitational fields.

Keywords: asteroid, boundary task, electric propulsion engine, gravitational potential, n-body problem, spacecraft, iESP, electrospray engine

1. Introduction

Over recent decades, there has been an increase in interest of the asteroids, comets, and other small celestial body studies. This is due to the changes in the research vector toward the solution of applied problems: counteraction to asteroids [1] and comet [2] hazards and development of small solar system objects with the purpose of mineral extraction [3, 4].

When designing research missions, there is a problem of the development of spacecraft control schemes in the attraction fields of complex geometric shapes bodies. The gravitational force between the elementary masses in that type of bodies is not sufficient to form the bodies in correct ellipsoidal or spheroidal form. Complex geometry creates the gravitational field of the complex configuration [1]. The spacecraft's behavior in this field is significantly different from the behavior of similar spacecraft near ellipsoidal and spheroidal bodies, the shape of which, as well as their gravitational field, can be considered correct in some approximation. The modeling of spacecraft motion and the development of control schemes near asteroids and comets is possible only under the condition that those bodies gravitational field formalization task was solved with the prescribed accuracy. However, if we take into consideration cavities and voids in the object structure, the center of masses displacement, and the uneven distribution of density [5], it becomes difficult to solve the gravitational field formalization problem.

In addition to the high accuracy of the gravitational field model, the lightweight of the model plays a huge role in ballistic design. The task of the spacecraft motion modeling and the task of finding optimally sustainable control schemes are virtually non-solvable because of the overloaded model of gravitational potential. That is why the task is to find a balance between the accuracy of the object's potential appearance and the convergence property of the task.

The problem of finding the correct mathematical description of the gravitational fields of different shaped objects was addressed in a number of sources. For example, in the paper [6], the authors address the polygonal method of presenting the gravitational potential of asteroid 4769 Castalia as a formal model. A comparison of the proposed approach with the model of point attracting center takes place in the paper [7], where authors present a comparative analysis of the polygonal model and the model of point attracting center for the asteroid 216 Cleopatra. The paper [8] described and compared the gravitational potential presented in the form of series expanded into spherical, spheroidal, and ellipsoidal functions for Martian moons. In the paper [9], the authors address the position of equilibrium points for 23 different asteroids in their polygonal models of gravitational fields.

The drawbacks of the approaches addressed are their cumbersomeness when using in the problems of spacecraft flight dynamics and the necessity to know in advance the physical properties of the objects—geometry and mass distribution. Modeling the gravitational field of objects whose properties and structure are not known in advance within the models described is not possible.

This work analyzes the possibility of using ion electrospray propulsions to stabilize the orbit of a spacecraft that maneuvers in the gravitational field of an object of complex geometric shape using the asteroid 433 Eros and 951 Gaspra as examples. The choice of these asteroids is due to the presence of sufficient information about the gravitational fields of these bodies [10–12]. Besides, 433 Eros and 951 Gaspra are quite massive, and therefore it is impossible to ignore the uneven distribution of their gravitational field in space [12–14]. The authors used this information in [1, 15] to prove the correctness of the models used.

As a result of motion simulation, the authors developed an idea of using the iESP module for the controlled motion in the vicinity of the asteroid. Such modules are able to gain necessary thrust and exhaust velocity for the stable motion near the asteroid.

2. Models of an irregular gravitational field

The chapter presents a model of the gravitational potential of asteroid 433 Eros and 951 Gaspra in terms of the applicability of the model in the problems of flight dynamics. It should be noted that the described model of motion does not require fundamental knowledge about the structure and composition of the object. The purpose of the analysis is to show that this method of describing the gravitational potential provides sufficient accuracy for subsequent flight design.

This work uses the asteroids 433 Eros and 951 Gaspra as examples cosmic bodies with irregular gravitational fields. The choices of these asteroids are explained by the shape of the asteroid, its mass and its orbit. According to [14], the physical properties and orbital characteristics asteroids are presented in **Table 1**.

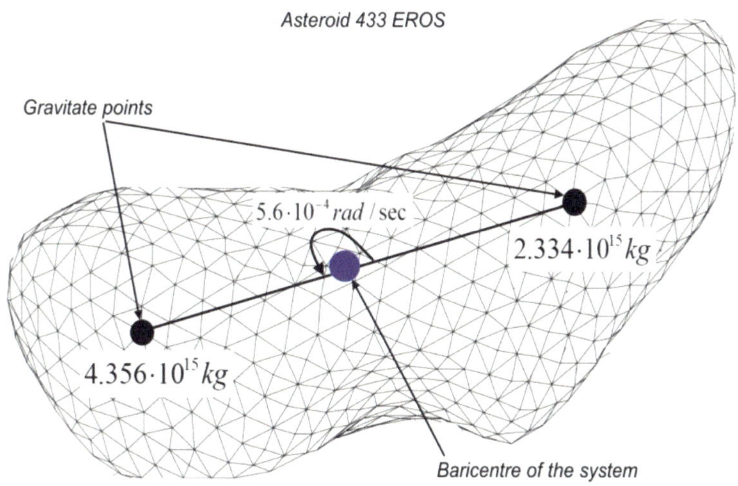

Figure 1.
Asteroid Eros 433 as a superposition of two rotation mass points.

A number of mathematical methods can be used for deriving a model of an irregular gravitational field. This chapter suggests the model of gravitational potential based on the n-body problem. Authors developed the idea of superposition of single gravitate mass points potential. The problem is to determine the necessary number of mass points and to define their locations. It is considered that two gravity points suffice for the simulation of gravitational field. It should be noted that the model under consideration assume that the asteroid's density is distributed evenly throughout the body to simplify the calculations.

In the paper [15], authors compare the barycentrical form of gravitational potential and gravitational potential in spherical harmonic coefficients. It is proved that barycentric potential is accurate enough for the motion simulations.

Let M_1, M_2 is masses of gravity points and $\mathbf{r}_1 = (x_1, y_1, z_1)$, $\mathbf{r}_2 = (x_2, y_2, z_2)$ are their vector positions (**Figure 1**). The corresponding form of the gravitational potential $U_{x,y,z}$ for this case is taken from [16]:

Physical characteristics of the asteroids 433 Eros and 951 Gaspra		
Mass, kg	6.69×10^{15}	$2\text{–}3 \times 10^{16}$
Density, kg/cm^3	2.67	2.7
Dimensions, km	$34.4 \times 11.2 \times 11.2$	$18.2 \times 10.5 \times 8.9$
Mean diameter, km	16.84	12.2
Rotation period, hours	5.27	7.04
Orbital characteristics of the asteroids 433 Eros and 951 Gaspra		
Eccentricity	0.22263	0.17363
Major axis, a.u.	1.45796	2.20971
Perihelion argument, deg	178.79567	129.72454
Inclination of orbit, deg	10.8289	4.10220
Circulation period, days	642.989	1199.782
Longitude of ascending node, deg	304.33453	253.17193

Table 1.
Physical and orbital characteristics of the asteroids 433 Eros and 951 Gaspra.

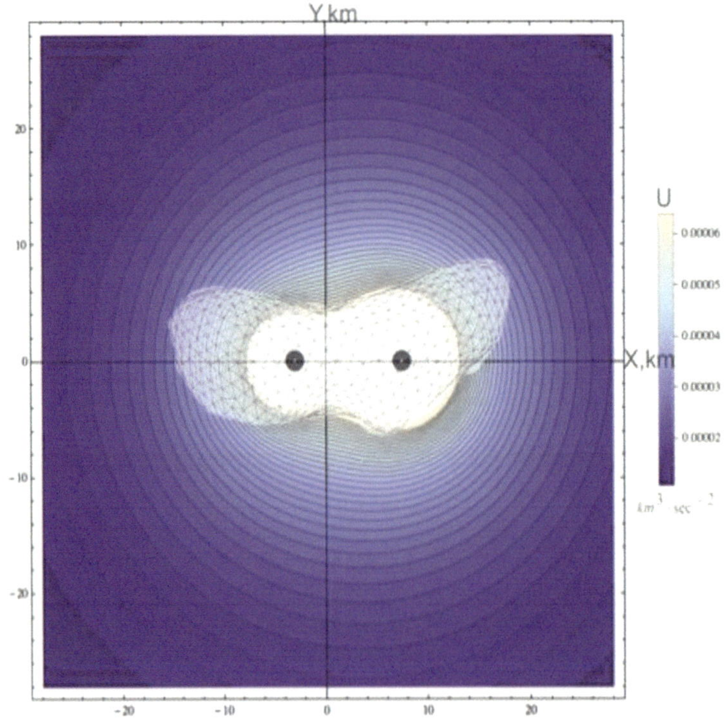

Figure 2.
Gravitational force lines of the Eros.

$$U_{x,y,z} = G \cdot \sum_{i=1}^{K=2} \frac{M_i}{\sqrt{(x-x_i)^2 + (y-y_i)^2 + (z-z_i)^2}} \tag{1}$$

$$\omega = \sqrt{\frac{G(M_1 + M_2)}{d^3}} \tag{2}$$

where G is the gravitational constant; $\mathbf{r} = (x, y, z)$ is vector of a spacecraft's position; ω is the angular velocity of the rotating around the barycentre; and d is the distance between the mass points. Hereafter, we call the potential (1) as the barycentric potential.

For example, in [15] we have analyzed the gravitational field of the asteroid 433 Eros and have calculated the exact mass values and the distance between the gravity points by trial and errors method. The following parameters of model $M_1 = 4.356 \cdot 10^{15}$ kg, $M_2 = 2.334 \cdot 10^{15}$ kg, $d = 10.8$ km, and $\omega = 5.6 \cdot 10^{-4}$ rad/s provide a minimum error in comparison with the gravitational potential described in the paper [11]. **Figure 2** shows the corresponding graph of the gravitational potential force lines for the Eros.

3. Motion simulation in an irregular gravitational field for an electric propulsion spacecraft

3.1 Simulation a passive spacecraft's motion

As we have already mentioned, the gravitational potential of the object can be represented as a set of attracting point masses. We will address the gravitational potential (1) in the inertial Cartesian three-dimensional coordinate system, which begins at the mass center *Bs* (**Figure 3**). The axis *BsZ* associates with the axis of asteroid's rotation.

We used the following assumptions to the simulations:

- the asteroid and the spacecraft are moving in the plane *XBsY*, the asteroid rotates clockwise, and the spacecraft rotates counter clockwise;
- the spacecraft has no impact on mass points motions (restricted circular three body problem);
- the Sun's gravity has impacted to the spacecraft and asteroid [17].

Vector form of the differential equations of the disturbed spacecraft's motion is [16]:

$$\frac{d^2\mathbf{r}}{dt^2} = -G\sum_{i=1}^{n}\frac{M_i}{\mathbf{r}_i^{\,3}}(\mathbf{r}-\mathbf{r}_i) + \mu_{Sun}\left(\frac{\Delta}{\Delta^3} - \frac{\mathbf{r}}{r^3}\right) + \frac{\mathbf{P}}{m_{SV}}; \tag{3}$$

$$\frac{dm_{SV}}{dt} = -\delta\tau \tag{4}$$

where $\mathbf{P}$ is controlling thrust vector; τ is propellant consumption per second; δ is Boolean-function of the engine operating condition (it can be 0 or 1); and m_{SV} is the spacecraft mass, μ_{Sun} is gravity parameter of Sun.

From now on, a radius is in km, velocity is in kilometer per second, thrust is in Newton, and angle is in radians. The integration scheme is the Runge-Kutta method (fourth-order accuracy). It is also noteworthy that all osculating parameters of orbit were obtained in the presumption that the asteroid's mass localizes in barycentre point.

We used the spacecraft's parameters, like to the missions Dawn and Rosetta [18], to simulate motion in these cases. The spacecraft's mass is 1200 kg and the exhaust velocity of the engine is 20 km/s. The initial commencing speed in every case was

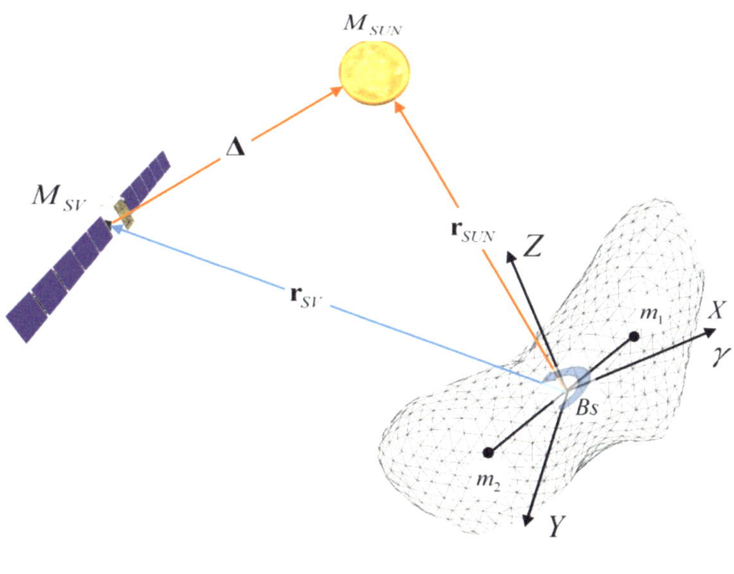

Figure 3.
Scheme of the three body problem.

chosen close to the circular velocity of the asteroid as the body with only one gravity point. The launch date is May 30, 2018. The authors developed the software for simulating spacecraft motion in irregular and regular gravitational fields [1].

To compare the impact of the distance from asteroid Eros on the spacecraft trajectory, we proposed to perform motion simulation for two orbit altitudes—50 km and 100 km. The same simulation to the asteroid Gaspra was performed for an orbit altitude of 200 km. **Table 2** presents the initial conditions for the three cases.

Figures 4 and **5** illustrate the trajectories of the passive spacecraft motion for two orbit altitudes in the plane: 50 km and 100 km from asteroid Eros. According to the graphs, orbits are not stable; moreover, in the first case, the spacecraft hits the asteroid after 18.35 days.

3.2 Simulation controlled spacecraft's motion: locally optimal controlled law

One of the problems of controlled spacecraft motion simulation is to find simple schemes to control the spacecraft that allow performing some simple maneuvers or stabilizing an orbit.

Case	r, km	V, km/s
Eros1	50	2.98×10^{-3}
Eros2	100	2.1×10^{-3}
Gaspra1	100	2.76×10^{-3}

Table 2.
Initial conditions for the three cases of simulation.

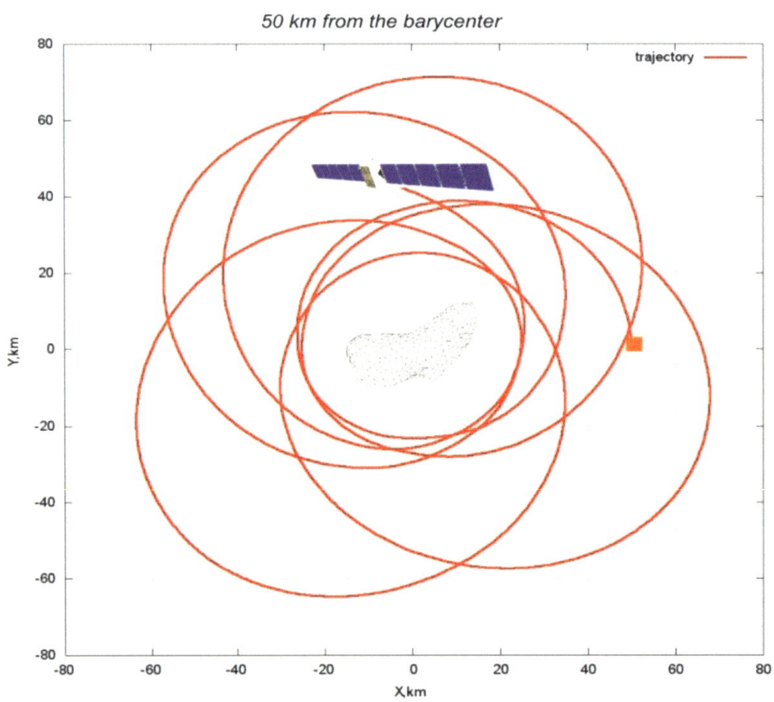

Figure 4.
Trajectory of passive spacecraft motion near the Eros in the plane XBsY (case Eros1)
$\boldsymbol{r} = (0 \quad 50 \quad 0)^T; \boldsymbol{V} = (0 \quad 2.98 \cdot 10^{-3} \quad 0)^T$.

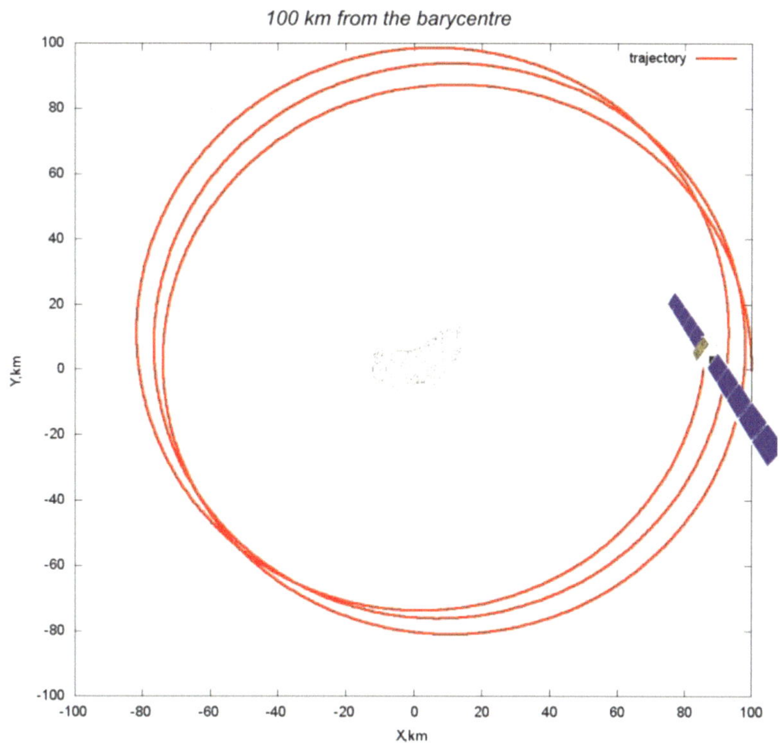

Figure 5.
Trajectory of passive spacecraft motion near the Eros in the plane XBsY (case Eros2) $\boldsymbol{r} = (0 \quad 100 \quad 0)^T$; $\boldsymbol{V} = (0 \quad 2.1 \cdot 10^{-3} \quad 0)^T$.

The primary requirement for these control schemes is the simplicity in operation. Consequently, it is suggested to use locally optimal actions based on osculating orbital elements. A spacecraft is assumed to move along the barycentric orbit, and corresponding osculating elements appertain to this orbit's plane. According to [17], application of the laws based on osculating orbital elements in this particular case allows obtaining a stable orbit for a spacecraft with an electric propulsion engine. It is suggested to use an angle η between the radius vector of the spacecraft $\mathbf{r}$ and the propulsive force vector $\mathbf{P}$ (**Figure 6**).

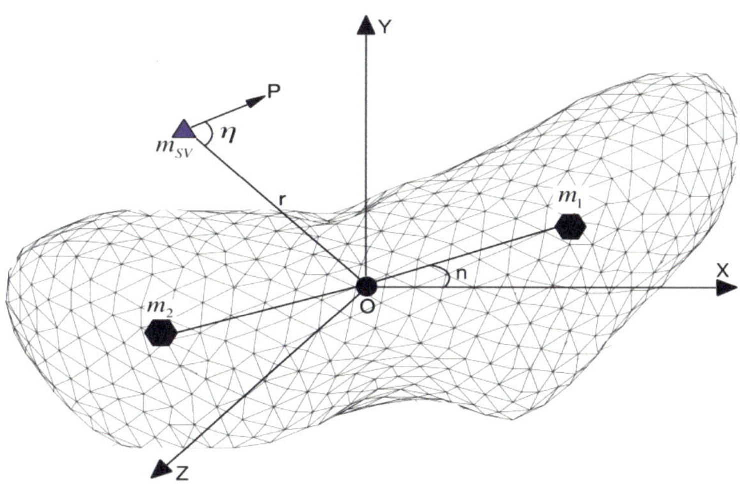

Figure 6.
Barycentric rectangular coordinate system—the Eros and the spacecraft appertain to the plane.

The constant semi-major axis law is considered [17] as the control law:

$$tg\eta = -\frac{e\sin\vartheta}{1+e\cos\vartheta} \tag{5}$$

where ϑ is true anomaly and e is eccentricity of the barycentric orbit.
The equations of the control thrust vector are:

$$P_x = \frac{r_x}{\sqrt{{r_x}^2+{r_y}^2}}\cos\eta - \frac{r_y}{\sqrt{{r_x}^2+{r_y}^2}}\sin\eta \tag{6}$$

$$P_y = \frac{r_y}{\sqrt{{r_x}^2+{r_y}^2}}\cos\eta + \frac{r_x}{\sqrt{{r_x}^2+{r_y}^2}}\sin\eta \tag{7}$$

$$P_z = 0 \tag{8}$$

The simulation parameters were determinated earlier. Additionally, the specific impulse of the engine module is IsP1 = 1000 s, IsP2 = 1500 s, and IsP3 = 2000 s. The trust values are various for the particular altitude. The particular values were obtained by the direct-search method.

The interpolated values of thrust levels for the orbit's stabilization control scheme (16) for IsP1 = 1000 s, IsP2 = 1500 s, and IsP3 = 2000 s are represented in **Figure** 7.

Figures 8 and **9** show graphs of changes in orbit parameters in case Eros1 and the engines with IsP1 = 1500 s.

Figure 10 shows the trajectory of spacecraft's motion in the OXY for 50 km Eros' orbit (case Eros1). The characteristic of thrust for the orbit altitude in 50 km and for the orbit altitude of 100 km are shown in **Table 3**.

Figure 11 shows trajectory of spacecraft's motion in the OXY for 100 km Eros' orbit (case Eros2).

Consider the trajectory of motion for the case of asteroid Gaspra, obtained by using the locally optimal control law—the constancy of the distance of the semi major axis (**Figure 12**), as well as graphs of changes in the barycentric coordinates

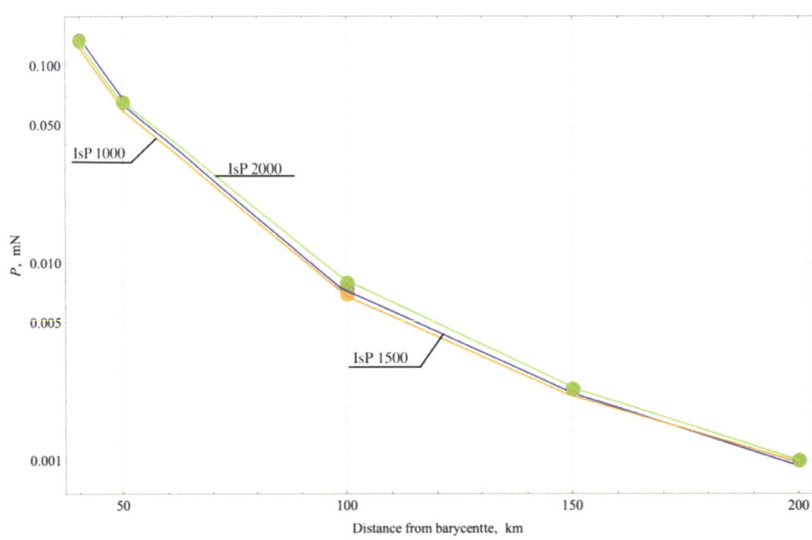

Figure 7.
Dependency of the thrust level to the barycentre distance and the specific impulse of propulsion system.

DOI: http://dx.doi.org/10.5772/intechopen.85615

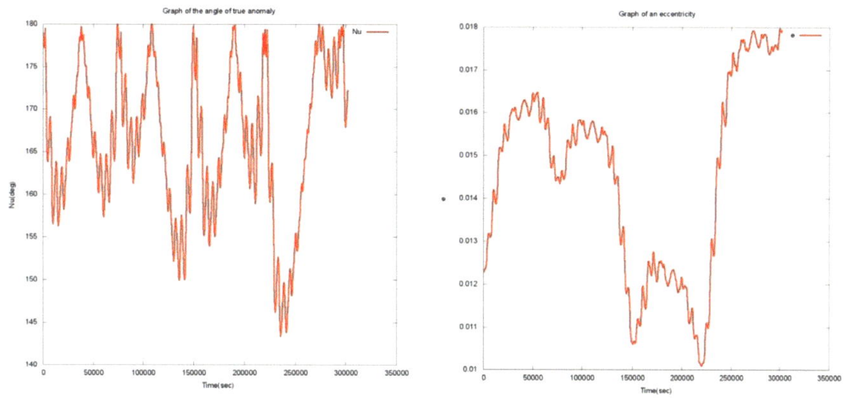

a) b)

Figure 8.
Graphs of changes in the parameters of the trajectory (case Eros1). (a) The angle of the true anomaly and (b) eccentricity.

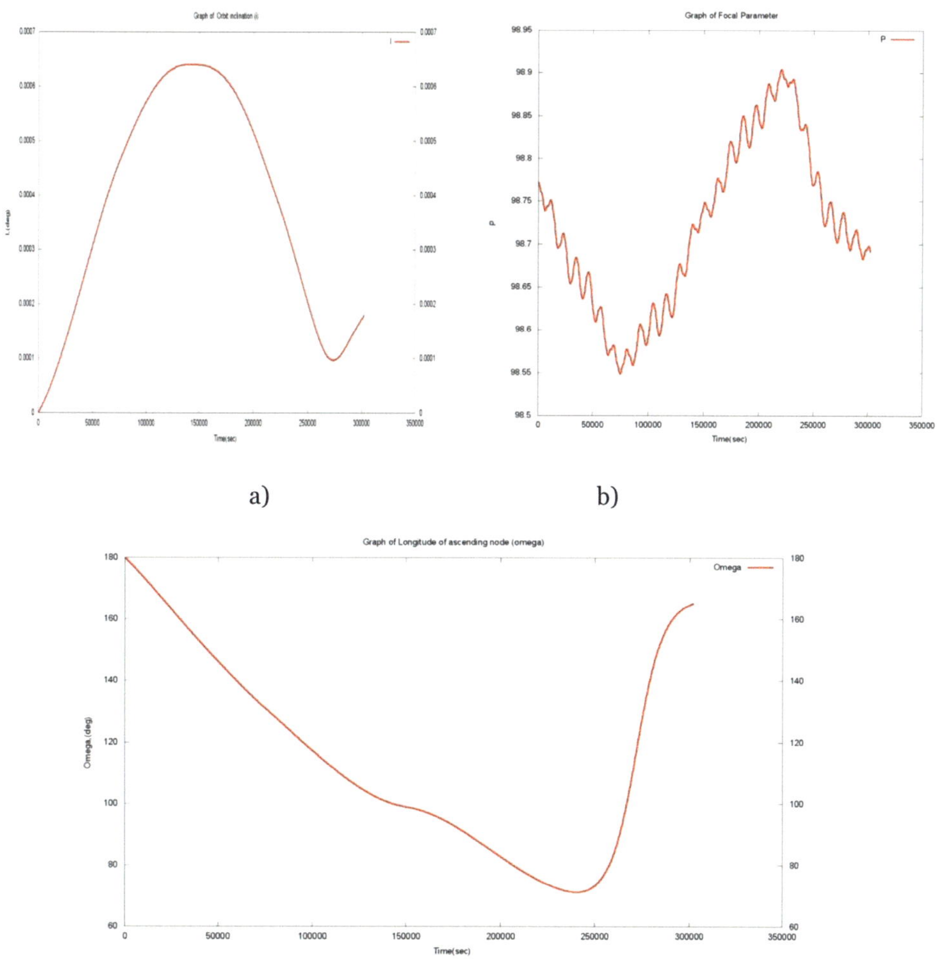

a) b)

c)

Figure 9.
Graphs of changes in the parameters of the trajectory (case Eros1). (a) Focal parameter, (b) inclination, and (c) longitude of the ascending node.

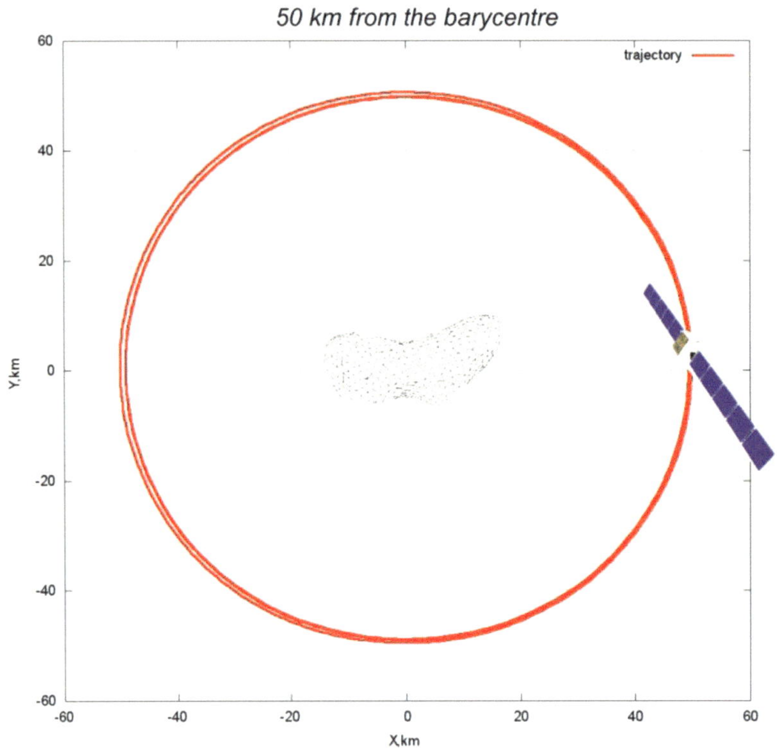

Figure 10.
Trajectory of spacecraft's motion near the Eros stabilized by the constant semi-major axis law $r = \begin{pmatrix} 0 & 50 & 0 \end{pmatrix}^T$; $V = \begin{pmatrix} 0 & 2.98 \cdot 10^{-3} & 0 \end{pmatrix}^T$, $P = 6.6 \cdot 10^{-5} N$ *(case Eros1).*

Case	r, km	P, N	c, km/s
Eros1	50	6.6×10^{-5}	15
Eros2	100	8×10^{-6}	15
Gaspra1	100	2.33×10^{-5}	20

Table 3.
Characteristics of thrust and exhaust velocity.

and velocities of the spacecraft (**Figures 13** and **14**) and graphs of changes in the orbit parameters (**Figures 15** and **16**).

4. Ion electrospray thruster assembly for spacecraft maneuvering in vicinity of asteroid

Ion electrospray engine (iESP) is a micro-thruster that is designed initially for CubeSats technology. Besides, such engines can be used for motion control systems of small spacecraft as these engines have a sufficient duration of operation and low consumption of the working fluid. It is a type of low thrust electric propulsion rocket engine that uses electrostatic acceleration of charged liquid droplets for propulsion. The liquid used for this application tends to be a low-volatility ionic liquid. The most amenable propellants for electrostatic thrusters have proven to be

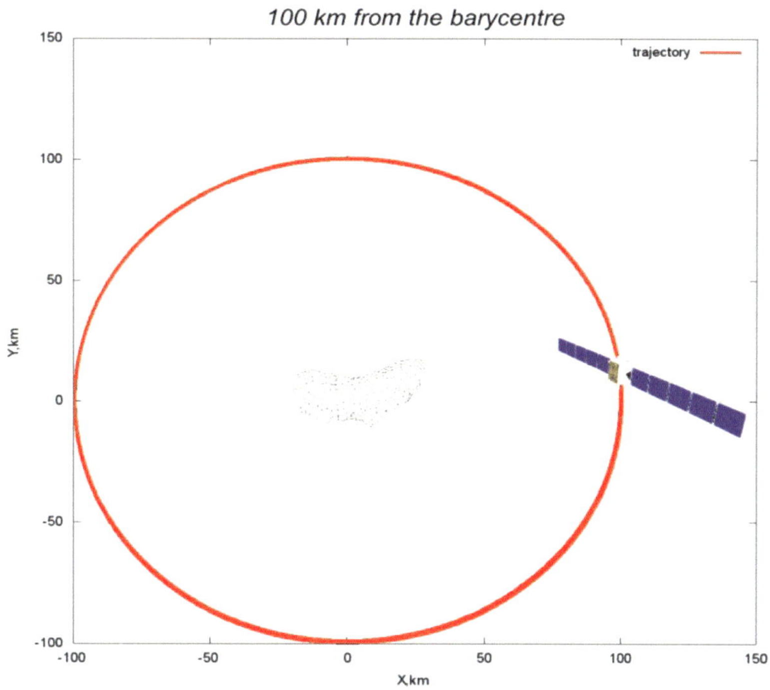

Figure 11.
Trajectory of spacecraft's motion near the Eros stabilized by the constant semi-major axis law $\boldsymbol{r} = (0 \quad 100 \quad 0)^T$; $\boldsymbol{V} = (0 \quad 2.1 \cdot 10^{-3} \quad 0)^T$, $P = 8 \cdot 10^{-6} N$ *(case Eros2).*

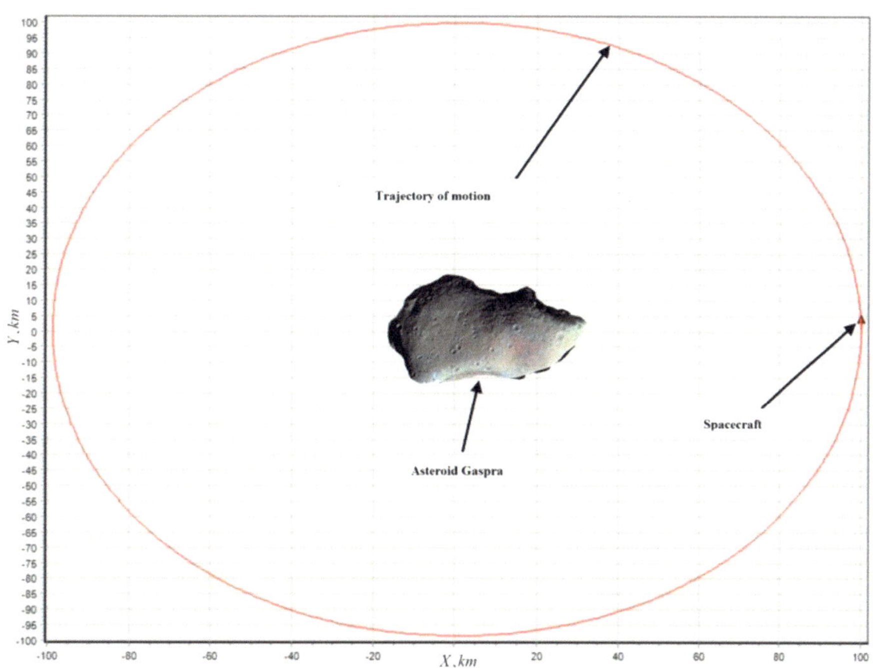

Figure 12.
The trajectory of the spacecraft for asteroid Gaspra (case Gaspra1).

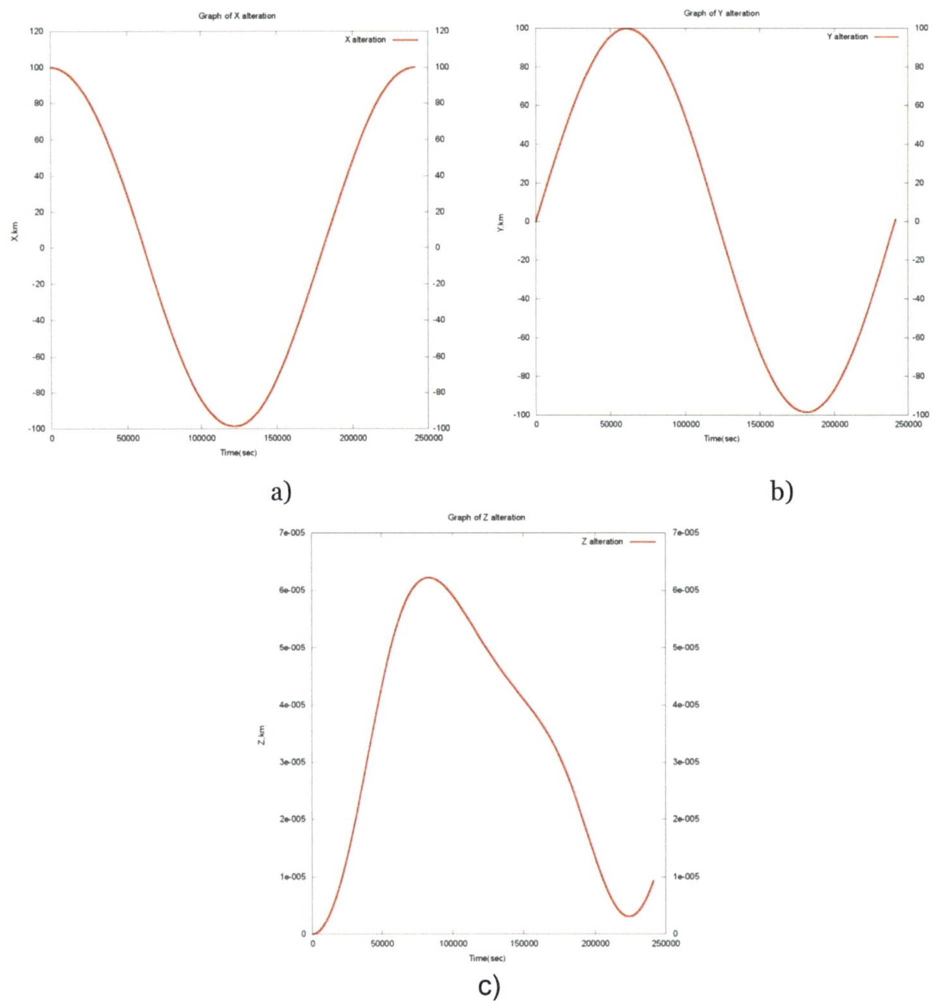

Figure 13.
Graphs of changes in the parameters of the trajectory (case Gaspra1). (a) X coordinate, (b) the Y coordinates, and (c) Z coordinates.

cesium, mercury, argon, krypton, and most commonly xenon, and many possible sources of such ions of the requisite efficiency, reliability, and uniformity have been conceived and developed [19].

The typical design appearances of an electrospray thruster are shown **Figures 17** and **18**.

Like other ion thrusters, the main benefits of iESP include high efficiency, thrust density, and specific impulse; however, it has shallow total thrust, on the order of micro-Newton. It provides excellent attitude control or dynamic acceleration of small spacecraft over long periods.

The spacecraft's motion simulation in asteroid's vicinity identifies that engine specification for the stabilization trajectories is close to the Ion electrospray thrusters' modules in the context of the thrust level and IsP. Therefore, we propose to use them not only as executive bodies of the control system, but also to stabilize the orbital motion of the spacecraft, which is unstable due to the disturbing action of an irregular gravitational field. The appearance of the control program identifies the number of demands for the electrospray engines module that is installed on the spacecraft. It can be used for the particular configuration of the power plant's design.

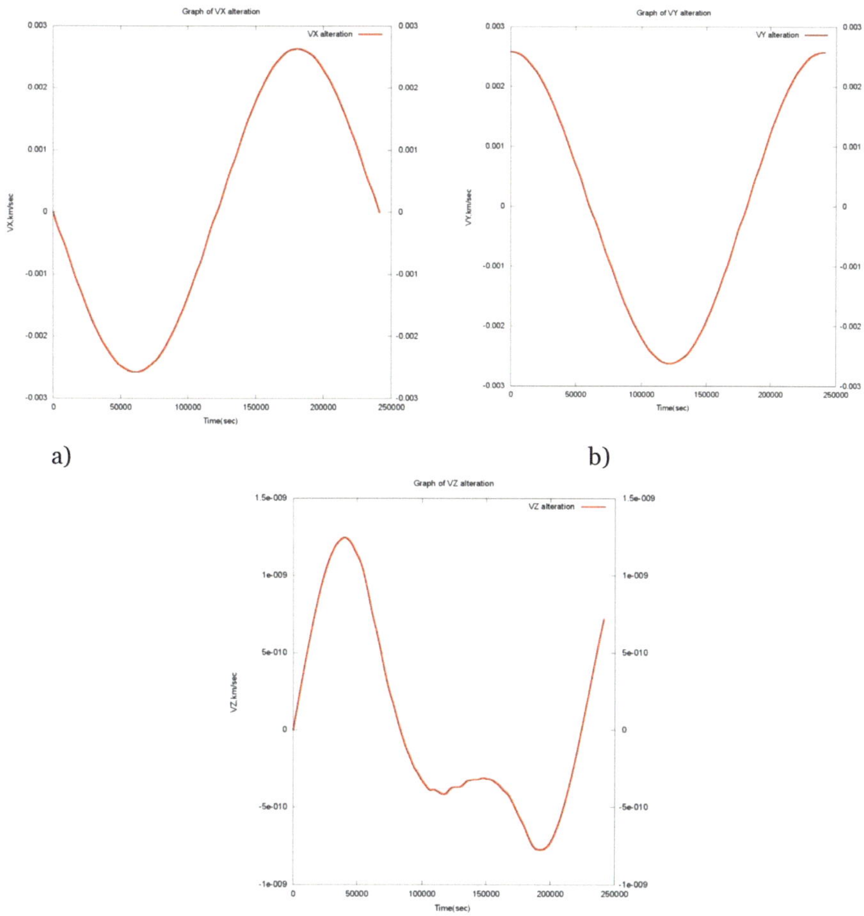

a) b) c)

Figure 14.
Graphs of changes in the parameters of the trajectory (case Gaspra1). (a) Vx velocity, (b) Vy velocity, and (c) Vz velocity.

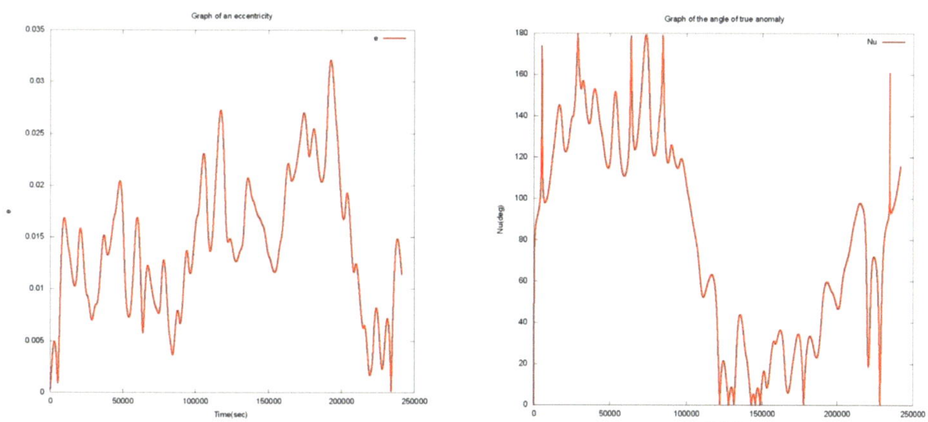

a) b)

Figure 15.
Graphs of changes in the parameters of the trajectory (case Gaspra1). (a) The eccentricity and (b) the angle of the true anomaly.

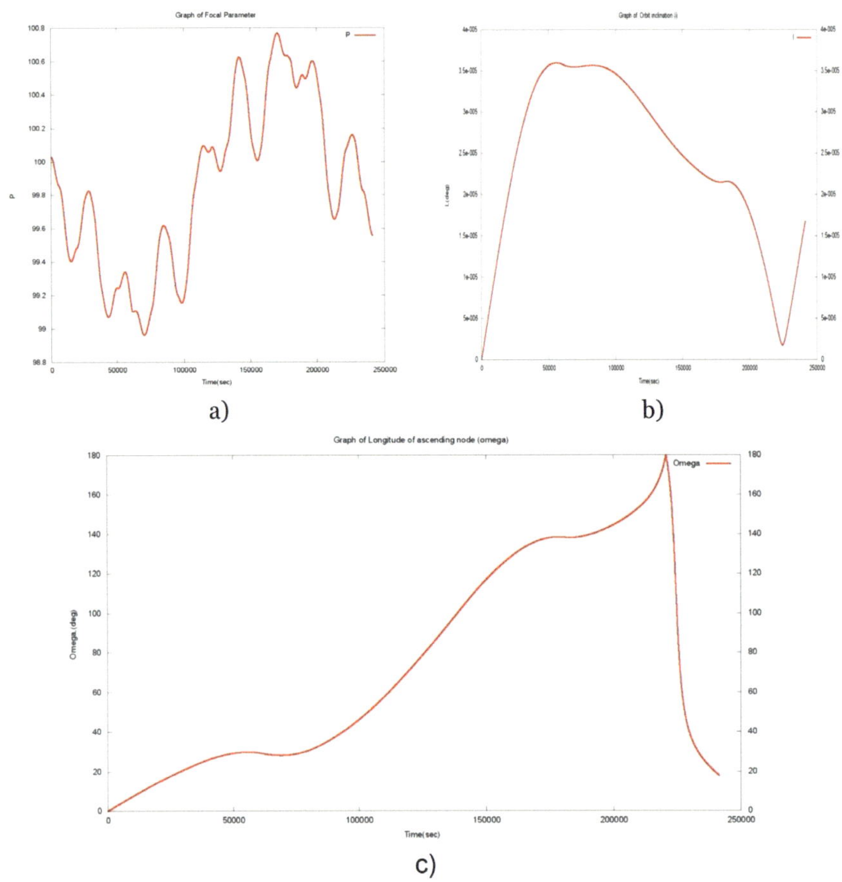

Figure 16.
Graphs of changes in the parameters of the trajectory (case Gaspra1). (a) Focal parameter, (b) inclination, and (c) longitude of the ascending node.

Figure 17.
Electrospray thruster.

Figure 18.
Electric engine.

Figures 19–21 show diagrams of the control law for the cases Eros1, Eros2, and Gaspra1 (control programs). It can be seen from the graphs that small changes in the direction of thrust of the engine (within 10°), according to the selected control law, allows making the flight orbits stable. Such control rules can be used to maneuver near an asteroid with an uneven gravitational field.

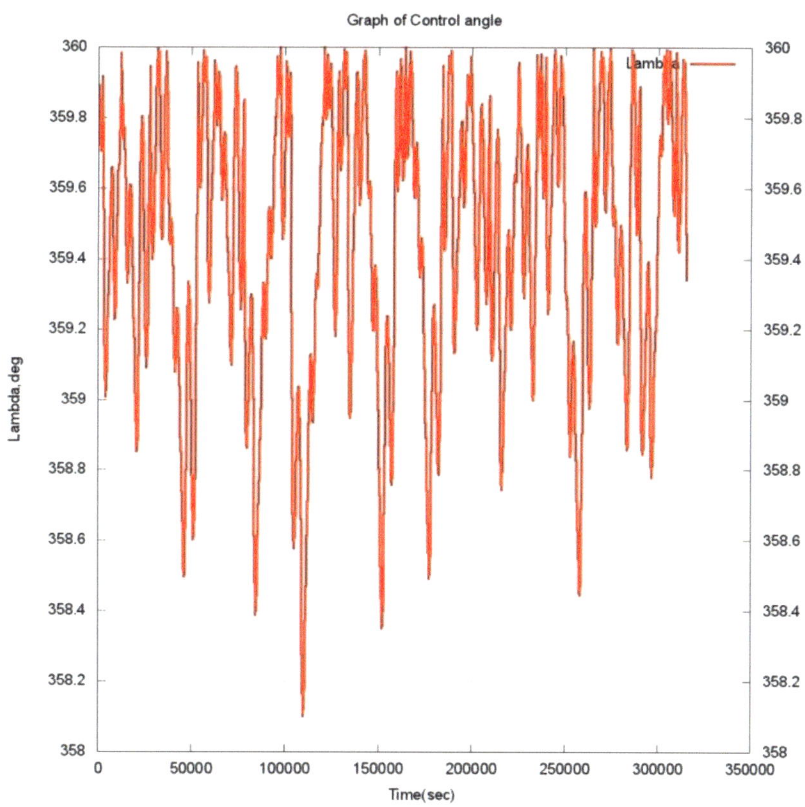

Figure 19.
The control program for the constant semi-major axis (50 km) (case Eros1).

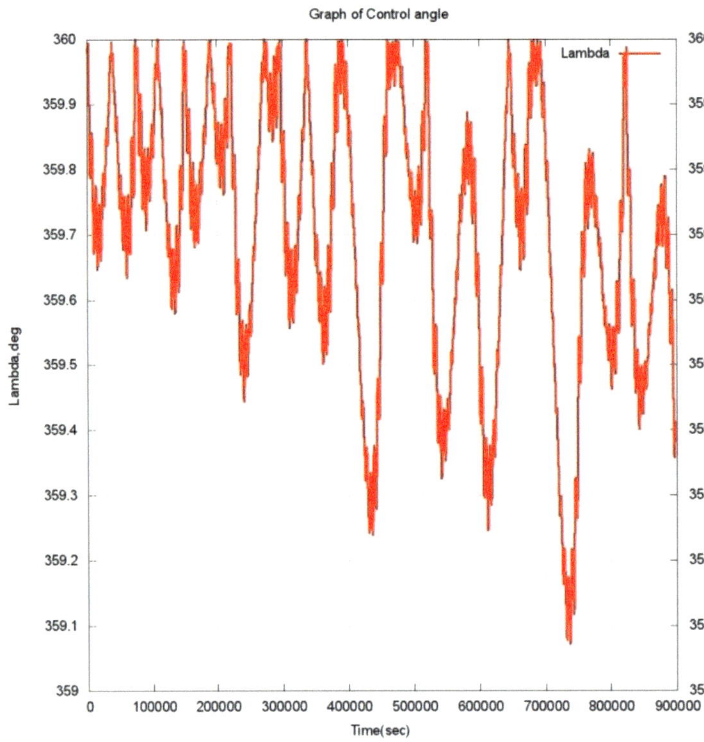

Figure 20.
The control program for the constant semi-major axis (100 km) (case Eros2).

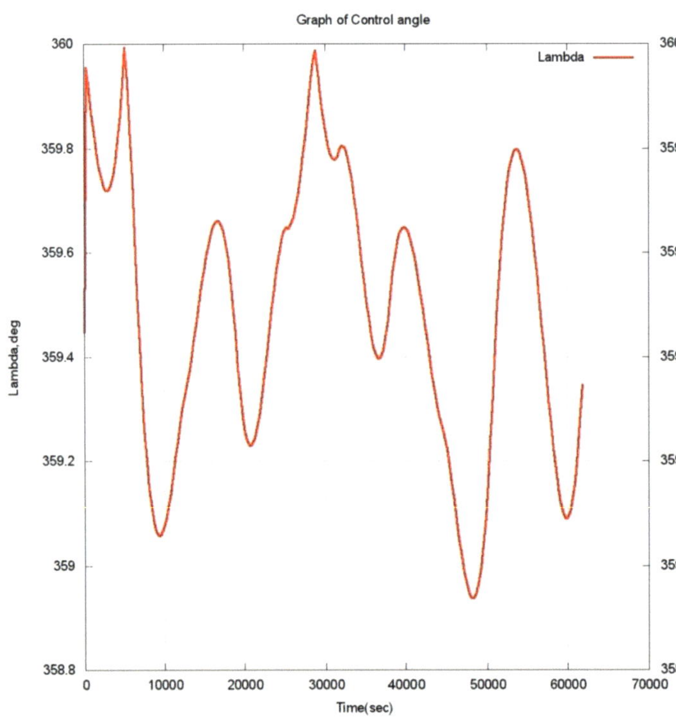

Figure 21.
The control program for the constant semi-major axis (100 km) (case Gaspra1).

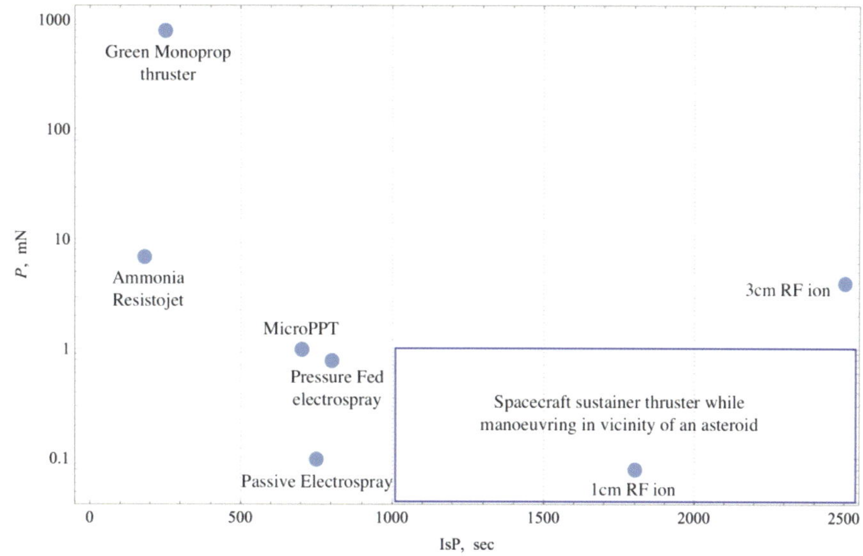

Figure 22.
Electric engine thrust/IsP distribution for suitable electric propulsion systems.

Thus, iESP modules can be successfully used not only for the control angular motion, but also for the movement in the vicinity of an asteroid in different research missions.

The simulation results of the orbital stabilization for other engines (for other specific impulse and thrust values) and orbit altitudes were not represented in the chapter; however, the obtained values were used for interpolation lines (**Figure 6**). Thrust levels were variable parameters that depend on the orbit altitude and the initial commencing speed. **Figure 22** shows the range of settings of electric propulsion systems suitable for use as an engine for maneuvering or stabilizing near-asteroid orbits.

5. Conclusions

This chapter discusses the use of electrospray engines to stabilize near-asteroid orbits. Presents two of the asteroids Eros and Gaspra have substantially nonspherical. Mathematical models of the asteroid's gravitational field and controlled motion of a spacecraft with an electric motor in an uneven gravitational field are developed.

The obtained results are used to simulate the passive and controlled motion of a spacecraft as an n-body problem. It is established that passive motion is unstable. The authors proposed to use the known locally optimal control law for orbit stabilization. A constant semi-axis control law of thrust was used to model the motion of three orbits (50, 100, and 200 km) near two asteroids (Eros and Gaspra) to stabilize. The levels of thrust and specific impulse and their dependence on the orbit height, initial velocity, and asteroid parameters were determined.

The simulation results show that the calculated thrust and specific impulse (or exhaust velocity) levels are within the range of ion motor characteristics. However, it is necessary to take into account the received control programs and the real and mass of the spacecraft. Therefore, under all these assumptions, it is possible to use the ion electrospray module for motion near asteroids.

The simulation results show that the calculated thrust and exhaust velocity levels are too low for the spacecraft's main engines. Thus, to guarantee the design characteristics, it is necessary to use an additional module iESP.

Acknowledgements

The authors state that part of the chapter was taken from our previously published article (Andrey Shornikov, Irina Gorbunova, Olga Starinova "Stabilized Trajectories of a Spacecraft in Inhomogeneous Gravitational Fields," AIP Publishing, 2018), and that we have the copyright to re-use it. The reported study was funded by Ministry of Education and Science of the Russian Federation according to the research project No. AAAA-A17-117031050032-9 (9.5453.2017).

Author details

Olga Starinova[1], Andrey Shornikov[1] and Elizaveta Nikolaeva[2]*

1 Department of Space Engineering, Samara National Research University, Samara, Russia

2 Interuniversity Department of Space Research, Samara National Research University, Samara, Russia

*Address all correspondence to: nikolaevalizaveta@mail.ru

References

[1] Shornikov A, Starinova O. Effectiveness analyses of an electrospray sustainer engine installed on a spacecraft maneuvering in vicinity of an asteroid. Procedia Engineering. 2017

[2] Chapman CR, Morrison D. On the Earth by asteroids and comet: Assessing the hazard. Nature. 1994;**367**:6458, 33-6440

[3] Ross SD. Near-Earth asteroid mining space. Use of near-Earth asteroids as platforms for future space bases; 2001

[4] Moore S. Technology development for NASA's asteroid redirect

[5] Britt DT et al. Asteroid density, porosity, and structure. In: Asteroids III. 1987

[6] Geissler P, Petit J-M, Durda D, Greenberg R, Bottke W, Nolan M, et al. Icarus. 1996;**120**:140

[7] Ren Y, Shan J. On tethered sample and mooring systems near irregular asteroids. Advances in Space Research. 2014;**54**(8):1608-1618

[8] Hu X, Jekeli C. A numerical comparison of spherical, spheroidal and ellipsoidal harmonic gravitational field models for small non-spherical bodies: Examples for the Martian moons. Journal of Geodesy. 2015;**89**(2):159-177

[9] Wang X, Jiang Y, Gong S. Analysis of the potential field and equilibrium points of irregular-shaped minor celestial bodies. Astrophysics and Space Science. 2014;**353**(1):105-121

[10] Michel P, Farinella P, Froeschlé C. The orbital evolution of the asteroid Eros and implications for collision with the Earth. Nature. 1996;**380**(6576):689

[11] Garmier R et al. Modeling of the Eros gravity field as is ellipsoidal harmonic expansion from the near Doppler track data. Geophysical Research Letters. 2002;**29**(8)

[12] Miller JK et al. Determination of shape, gravity, and rotational state of asteroid 433 Eros. Icarus. 2002;**155**(1): 3-17

[13] Moore C. Technology Development for NASA's Asteroid Redirect Mission, IAC-14-D2.8-A5.4.1. Available from: https://www.nasa.gov/sites/default/iles/iles/IAC-14-D2_8-A5_4_1-Moore.pdf

[14] Asteroids internet base. Available from: http://space.frieger.com/asteroids/

[15] Shornikov A, Starinova O. Simulation of controlled motion in an irregular gravitational field for an electric propulsion spacecraft. In: Proceedings of the IEEE 7th International Conference on Recent Advances in Space Technologies (RAST); 16–19 June 2015; Istanbul, Turkey. 2015. pp. 771-776

[16] Szebehely V. Theory of Orbits: The Restricted Problem of Three Bodies. New Haven, CT: Yale University; 1967. pp. 10-25

[17] Lebedev V. The Calculation of the Motion of a Spacecraft with Low Thrust. Moscow: Computation Centre of the Russian Academy of Sciences; 1968. pp. 4-10

[18] Scheeres D. The orbital dynamics environment of 433 Eros. Ann Arbor. 2002. 1001

[19] Sheth V. Spacecraft electric propulsion—A review. International Journal of Research in Aeronautical and Mechanical Engineering. 2014;43-55. ISSN: 2321-3051